A Student's Guide to Waves

Waves are an important topic in the fields of mechanics, electromagnetism, and quantum theory, but many students struggle with the mathematical aspects. Written to complement course textbooks, this book focuses on the topics that students find most difficult.

Retaining the highly popular approach used in Fleisch's other Student's Guides, the book uses plain language to explain fundamental ideas in a simple and clear way. Exercises and fully worked examples help readers test their understanding of the concepts, making this an ideal book for undergraduates in physics and engineering trying to get to grips with this challenging subject.

The book is supported by a suite of online resources available at www.cambridge.org/wavesguide. These include interactive solutions for every exercise and problem in the text and a series of podcasts in which the authors explain the important concepts of every section of the book.

DANIEL FLEISCH is a Professor in the Department of Physics at Wittenberg University, where he specializes in electromagnetics and space physics. He is the author of several Student's Guide books, including most recently *A Student's Guide to the Mathematics of Astronomy* (Cambridge University Press, 2013).

LAURA KINNAMAN is an Assistant Professor of Physics at Morningside College, where she carries out computational research in chemical physics and organizes the Physics Club.

A Student's Guide to Waves

DANIEL FLEISCH
Wittenberg University

LAURA KINNAMAN
Morningside College

CAMBRIDGE
UNIVERSITY PRESS

University Printing House, Cambridge CB2 8BS, United Kingdom

Cambridge University Press is part of the University of Cambridge.

It furthers the University's mission by disseminating knowledge in the pursuit of education, learning and research at the highest international levels of excellence.

www.cambridge.org
Information on this title: www.cambridge.org/9781107054868

First published 2015

A catalogue record for this publication is available from the British Library

Library of Congress Cataloguing in Publication data
Fleisch, Daniel A., author.
A student's guide to waves / Daniel Fleisch, Wittenberg University, Laura Kinnaman, Morningside College.
pages cm
Includes bibliographical references and index.
ISBN 978-1-107-05486-8 (Hardback) – ISBN 978-1-107-64326-0 (Paperback)
1. Waves–Textbooks. I. Kinnaman, Laura, author. II. Title.
QC157.F54 2015
530.12'4–dc23 2014032243

ISBN 978-1-107-05486-8 Hardback
ISBN 978-1-107-64326-0 Paperback

Additional resources for this publication at www.cambridge.org/9781107643260

Contents

Preface

This book has one purpose: to help you understand the foundational concepts of waves and the mathematics of the wave equation. The authors have attempted to fill the book with clear, plain-language explanations, using just enough mathematical rigor to help you understand the important principles without obscuring the underlying physics. Armed with that understanding, you'll be ready to tackle the many excellent texts that deal with mechanical, electromagnetic, and quantum waves.

You should understand that this book is meant to be used as a supplemental text and is not intended to be a comprehensive treatment of wave phenomena. That means that we haven't attempted to cover every aspect of waves; instead, we've included the topics that our students have found most troubling.

As you'll see, the design of the book supports its use as a supplemental text. Whenever possible, we've made the chapters modular, allowing you to skip material you've already mastered so you can proceed directly to the topics with which you need help. As a Student's Guide, this book is accompanied by a website that provides a variety of freely available material that we think you'll find very helpful. That includes complete, interactive solutions to every problem in the book, as well as a series of podcasts in which we explain the most important concepts, equations, and graphs in every section of every chapter. By "interactive" we mean that you can see the full solution immediately, or you can request one or more hints that will guide you to the final solution. The icon appears throughout the book and highlights where there is accompanying material available online. If you choose to read the ebook on a device that supports interactivity, these additional features will appear directly within the text. If your device doesn't support interactivity, clicking on will take you straight to the books website.

Is this book right for you? It is if you're looking for help in understanding waves, whether you need that help to supplement your work in a physics or engineering class, in preparing for the physical science portion of a standard exam, or as a part of a program of self-study. Whatever your reason, we commend your initiative.

Acknowledgements

Primary responsibility for the good bits in this book belongs to the students in our classes, whose curiosity, intelligence, and persistence have inspired us to pursue (and occasionally find) deeper understanding and better explanations of the physics of waves. We thank those students.

We also thank Dr. Nick Gibbons, Dr. Simon Capelin, and the world-class production team of Cambridge University Press, whose support has been essential during the two-year process that has resulted in this book. The e-book version of this text would not have been possible without the thoughtful guidance of Claire Eudall and Catherine Flack.

Laura also thanks her sister, Dr. Carrie Miller, for all of the feedback, support, and encouragement that she's given. I can always count on Carrie to help me see my way out of a tricky spot. I also thank Bennett for his patience and support when I hole myself up, writing. My parents, sisters, brothers-in-law, nieces, and nephews who all provided encouragement and diversion, thank you!

And as always, Dan thanks Jill for her unwavering support and acknowledges the foresight and intuition of Dr. John Fowler, who made possible my contributions to the Cambridge line of *Student's Guides*.

1
Wave fundamentals

This chapter covers the fundamental concepts of waves. As with all the chapters in the book, you can read the sections within this chapter in any order, or you can skip them entirely if you're already comfortable with this material. But if you're working through one of the later chapters and you find that you're uncertain about some aspect of the discussion, you can turn back to the relevant section of this chapter.

In the first two sections of this chapter you'll be able to review the basic definitions and terminology of waves (Section 1.1) and the relationships between wave parameters (Section 1.2). Later sections cover topics that serve as the foundation on which you can build your understanding of waves, including vectors (Section 1.3), complex numbers (Section 1.4), the Euler relations (Section 1.5), wavefunctions (Section 1.6), and phasors (Section 1.7).

1.1 Definitions

When you're embarking on a study of new topic, it's always a good idea to make sure you understand the terminology used by people who discuss that topic. Since this book is all about waves, a reasonable place to start is by asking the question "What exactly is a wave?"

Here are some of the answers to that question that you may encounter in the literature.

"A classical traveling wave is a self-sustaining disturbance of a medium, which moves through space transporting energy and momentum." [6].

"What is required for a physical situation to be referred to as a wave is that its mathematical representation give rise to a partial differential equation of a particular form, known as *the wave equation*." [9].

"[The essential feature of wave motion is that a] condition of some kind is transmitted from one place to another by means of a medium, but the medium itself is not transported." [4].

"[A wave is] each of those rhythmic alternations of disturbance and recovery of configuration."[1]

Although there's not a great deal of commonality in these definitions of a wave, each contains an element that can be very helpful when you're trying to decide whether some phenomenon can (or should) be called a wave.

The most common defining characteristic is that a wave is a *disturbance* of some kind, that is, a change from the equilibrium (undisturbed) condition. A string wave disturbs the position of segments of the string, a sound wave disturbs the ambient pressure, an electromagnetic wave disturbs the strengths of the electric and magnetic fields, and matter waves disturb the probability that a particle exists in the vicinity.

In *propagating* or *traveling* waves, the wave disturbance must move from place to place, carrying energy with it. But you should be aware that combinations of propagating waves can produce non-propagating disturbances, such as those of a standing wave (you can read more about this in Section 3.2 of Chapter 3).

In *periodic* waves, the wave disturbance repeats itself in time and space. So, if you stay in one location and wait long enough, you're sure to see the same disturbance as you've seen previously. And if you take an instantaneous snapshot of the wave, you'll be able to find different locations with the same disturbance. But combinations of periodic waves can add up to non-periodic disturbances such as a wave pulse (which you can read about in Section 3.3 of Chapter 3).

Finally, in *harmonic* waves, the shape of the wave is sinusoidal, meaning that it takes the form of a sine or cosine function. You can see plots of a sinusoidal wave in space and time in Fig. 1.1.

So waves are disturbances that may or may not be propagating, periodic, and harmonic. But whatever the type of wave, there are a few basic parameters that you should make sure you understand. Here's a little FAQ that you may find helpful.

Q: How far is it from one crest to the next?

A: λ (Greek letter "lambda"), the **wavelength**. Wavelength is the amount of distance per cycle and has dimensions of length; in SI,[2] the units

[1] Oxford English Dictionary.

[2] "SI" stands for "Système International d'unités", the standard metric reference system of units.

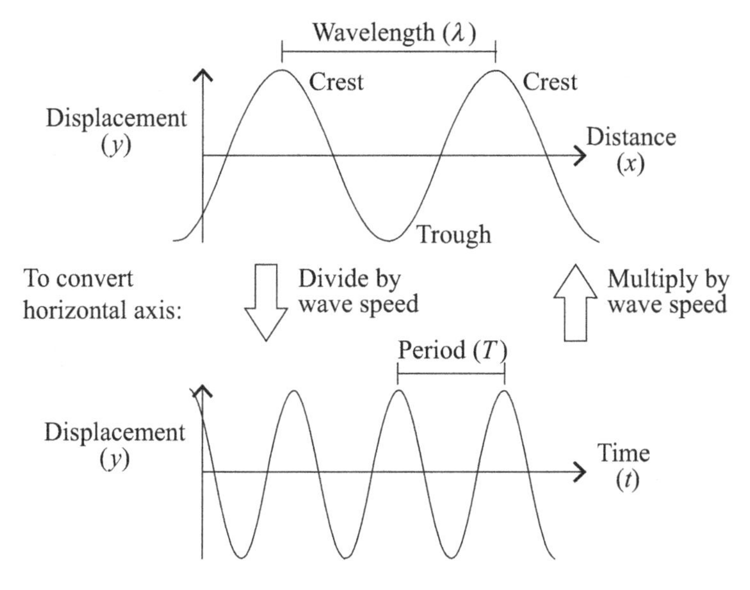

Figure 1.1 An example of a sinusoidal wave plotted in space and time.

of length are meters (m). Shouldn't this technically be "meters/cycle"? Yes, but since people know you're taking about waves when you mention wavelength, the "per cycle" is usually assumed and not explicitly kept in the units.

Q: How long in time is it between crests?

A: T (sometimes you'll see this written as P), the **period**. Period is the amount of time per cycle and has units of time, seconds (s) in SI. Again, this is really "seconds per cycle", but the "per cycle" is assumed and usually dropped.

Q: How often do crests come by?

A: f, the **frequency**. If you count how many wave crests pass by a given place in a certain amount of time, you are measuring f. Thus frequency is the number of cycles per amount of time and has units of one over time (technically cycles per unit time, but again, "cycles" is assumed and may be omitted). So in SI you'll see the units of frequency either as cycles/sec or $1/\text{s}$, which are also called hertz (Hz). The frequency of a wave is the inverse of the wave's period (T).

An illustration of the meaning of wavelength, wave period, and frequency (and how they're measured) is shown in Fig. 1.2.

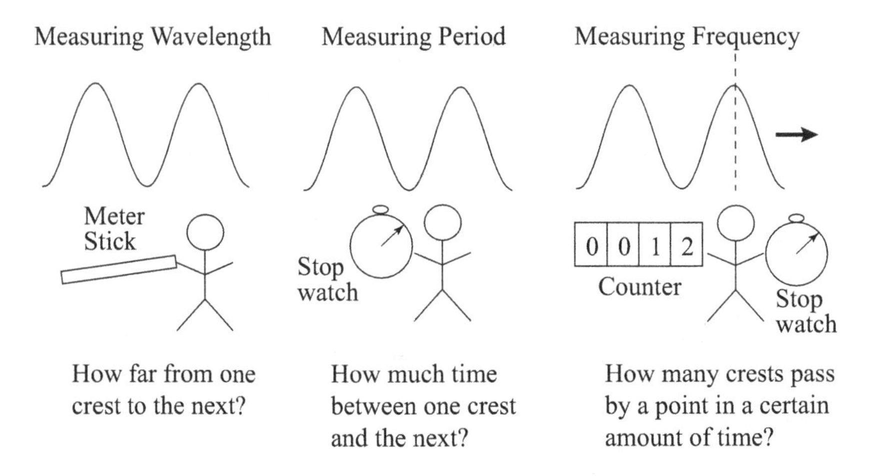

Figure 1.2 Measuring wave parameters.

Q: How big is the wave at any given place or time?

A: y, the **displacement**. Displacement is the amount of disturbance from equilibrium produced by the wave; its value depends on the place and time at which you measure the wave (and so is a function of x and t for a wave moving along the x-axis). The units of displacement depend on exactly what kind of a wave it is: waves on strings have displacements with units of distance (see Chapter 4), electromagnetic waves have displacements with units of electric and magnetic field strength (see Chapter 5), and one-dimensional quantum-mechanical matter waves have displacement with units of one over the square root of length (see Chapter 6).

Q: What is the biggest the wave ever gets?

A: A, the **amplitude**. Amplitude is a special value related to the displacement that occurs at the peak of a wave. We say "related to" because there are several different types of amplitude. "Peak" amplitude is the maximum displacement from equilibrium; this is measured from the equilibrium value to the top of the highest peak or the bottom of the deepest trough. "Peak-to-peak" amplitude is the difference between a positive peak and a negative peak, measured from crest to trough. And "rms" amplitude is the root-mean-square value of the displacement over one cycle. For sinusoidal waves, the peak-to-peak amplitude is twice as big as the peak amplitude, and the rms amplitude is 0.707 times the peak amplitude. Amplitude has the same units as displacement.

Q: How fast is the wave moving?

A: v, the **wave speed**. Usually, when authors refer to wave speed, they're talking about **phase speed**: How fast does a given point on a wave move? For example, if you measure how long it takes for one crest of a wave to travel a certain distance, you're measuring the phase speed of the wave. A different speed, **group speed**, is important for groups of waves called wave packets whose shape may change over time; you can read more about this in Section 3.4 of Chapter 3.

Q: What determines which part of a wave is at a given place at a certain time?

A: ϕ (Greek letter "phi"), the **phase**. If you specify a place and time, the phase of the wave tells you whether a crest, a trough, or something in between will appear at that place and time. In other words, phase is the argument of the function that describes the wave (such as $\sin\phi$ or $\cos\phi$). Phase has SI units of radians and values between 0 and $\pm 2\pi$ over one cycle (you may also see phase expressed in units of degrees, in which case one cycle $= 360° = 2\pi$ radians).

Q: What determines the starting point of a wave?

A: ϵ (Greek letter "epsilon"), or ϕ_0 ("phi-zero"), the **phase constant**. At the time $t = 0$ and location $x = 0$, the phase constant ϵ or ϕ_0 tells you the phase of the wave. If you encounter two waves that have the same wavelength, frequency, and speed but are "offset" from one another (that is, they don't reach a peak at the same place or time), those waves have different phase constants. A cosine wave, for example, is just a sine wave with a phase-constant difference of $\pi/2$, or 90°.

Q: All this sounds suspiciously like phase is related to some kind of angle.

A: That's not a question, but you're right, which is why phase is sometimes called "phase angle". The next two definitions should help you understand that.

Q: What relates a wave's frequency or period to angles?

A: ω (Greek letter "omega"), the **angular frequency**. The angular frequency tells you how much angle the phase of the wave advances in a given amount of time, so the SI units of angular frequency are radians per second. Angular frequency is related to frequency by the equation $\omega = 2\pi f$.

Q: What relates a wave's wavelength to angles?

A: k, the **wavenumber**. The wavenumber tells you how much the phase of the wave advances in a given amount of distance, so wavenumber has SI units of radians per meter. Wavenumber is related to wavelength by the equation $k = 2\pi/\lambda$.

1.2 Basic relationships

Many of the basic wave parameters defined in the previous section are related to one another through simple algebraic equations. For example, the frequency (f) and the period (T) are related by

$$f = \frac{1}{T}. \tag{1.1}$$

This equation tells you that frequency and period are *inversely* proportional. This means that longer period corresponds to lower frequency, and shorter period corresponds to higher frequency.

You can verify that Eq. (1.1) is dimensionally consistent by recalling from Section 1.1 that the units of frequency are cycles/second (often expressed simply as 1/s) and the units of period are just the inverse: seconds/cycle (usually expressed as "s"). So the dimensions of Eq. (1.1) in SI units are

$$\left[\frac{\text{cycles}}{\text{seconds}}\right] = \left[\frac{1}{\text{seconds/cycle}}\right].$$

Another simple but powerful equation relates the wavelength (λ) and frequency (f) of a wave to the wave's speed (v). That equation is

$$\lambda f = v. \tag{1.2}$$

The basis for this equation can be understood by considering the fact that speed equals distance divided by time, and a wave covers a distance of one wavelength in a time interval of one period. Hence $v = \lambda/T$, and since $T = 1/f$, this is the same as $v = \lambda f$. It also makes physical sense, as you can see by considering a wave that has long wavelength and high frequency. In that case, the speed of the wave must be high, for how else could those far-apart crests (long wavelength) be coming past very often (high frequency)? Now think about a wave for which the wavelength and frequency are both small. Since those closely spaced crests (short wavelength) are not coming past very often (low frequency), the wave must be moving slowly.

To see that the dimensions are balanced in Eq. (1.2), consider the units of wavelength multiplied by the units of frequency:

$$\left[\frac{\text{meters}}{\text{cycle}}\right]\left[\frac{\text{cycles}}{\text{second}}\right] = \left[\frac{\text{meters}}{\text{second}}\right],$$

which are the units of speed.

So Eq. (1.2) allows you to find the speed of a wave if you know the wave's wavelength and frequency. But, as you study waves, you're likely to encounter many situations in which you're dealing with waves of the same type that

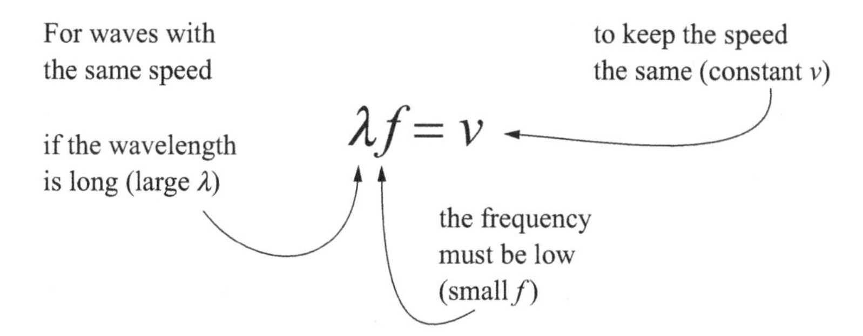

Figure 1.3 The relationship of wavelength to frequency for waves of the same speed.

are moving at the same speed (such as electromagnetic waves in a vacuum, which all travel at the speed of light). In such cases, the waves may have different wavelength (λ) and frequency (f), but the *product* of the wavelength and frequency must equal the wave speed.

This means that as long as the wave speed (v) is constant, waves with longer wavelength (large λ) must have lower frequency (small f). Likewise, for waves with the same speed, if the wavelength is short (small λ), the frequency must be high (large f). This concept is so important we've written it as an "expanded equation" in Fig. 1.3.

For sound waves (which have constant speed under certain circumstances), frequency corresponds to pitch. So low-pitch sounds (such as the bass notes of a tuba or the rumble of a passing truck) must have long wavelength, and high-pitch sounds (such as the tweets of a piccolo or Mickey Mouse's voice) must have short wavelength.

For electromagnetic waves in the visible portion of the spectrum, frequency corresponds to color. So the relationship between wavelength, frequency, and speed means that low-frequency (red) light has longer wavelength than high-frequency (blue) light.

There are two additional equations that are very useful when you're working on wave problems. The first of these is the relationship between frequency (f), period (T), and angular frequency (ω):

$$\omega = \frac{2\pi}{T} = 2\pi f. \tag{1.3}$$

You can see from this equation that angular frequency has dimensions of angle over time (SI units of rad/s), consistent with the definition of this parameter in Section 1.1. So frequency (f) tells you the number of cycles per second, and angular frequency (ω) tells you the number of radians per second.

Here's why the angular frequency (ω) of a wave is a useful parameter. Let's say you want to know how much the phase of a wave will change at a certain location in a given amount of time (Δt). To find that phase change ($\Delta\phi$), just multiply the angular frequency (ω) by the time interval (Δt):

$$(\Delta\phi)_{\text{constant } x} = \omega\,\Delta t = \left(\frac{2\pi}{T}\right)\Delta t = 2\pi\left(\frac{\Delta t}{T}\right), \tag{1.4}$$

where the subscript "constant x" is a reminder that this change in phase is due only to advancing time. If you change location, there will be an additional phase change as described below, but for now we're considering the phase change at one location (constant x).

At this point, it may help you to step back from Eq. (1.4) and take a look at the $\Delta t/T$ term. This ratio is just the fraction of a full period (T) that the time interval Δt represents. Since the phase change during a full period is 2π radians, multiplying this fraction ($\Delta t/T$) by 2π radians gives you the number of radians that the wave phase has advanced during the time interval Δt.

Example 1.1 *How much does the phase of a wave with period (T) of 20 seconds change in 5 seconds?*

Since the wave period T is 20 seconds, a time interval Δt of 5 seconds represents 1/4 period ($\Delta t/T = 5/20 = 1/4$). Multiplying this fraction by 2π gives $\pi/2$ radians. Thus the phase of the wave advances by $\pi/2$ radians (90°) every 5 seconds.

This illustrates why angular frequency (ω) can be thought of as a "time-to-phase converter". Given any amount of time t, you can convert that time to phase change by finding the product ωt.

The final important relationship of this section concerns wavenumber (k) and wavelength (λ). The relationship between these parameters is

$$k = \frac{2\pi}{\lambda}. \tag{1.5}$$

This equation shows that wavenumber has the dimensions of angle over distance (with SI units of rad/m). It also suggests that wavenumber can be used to convert distance to phase change, just as angular frequency can be used to convert time to phase change.

To find the phase change $\Delta\phi$ over a given distance at a certain time, multiply the wavenumber k by a distance interval Δx:

$$(\Delta\phi)_{\text{constant } t} = k\,\Delta x = \left(\frac{2\pi}{\lambda}\right)\Delta x = 2\pi\left(\frac{\Delta x}{\lambda}\right), \tag{1.6}$$

where the subscript "constant t" is a reminder that this change in phase is due only to changing location (as described above, there will be an additional phase change due to the passage of time).

Just as the term $\Delta t/T$ gives the fraction of a full cycle represented by the time interval Δt, the term $\Delta x/\lambda$ gives the fraction of a full cycle represented by the distance interval Δx. Thus the wavenumber k serves as a "distance-to-phase converter", allowing you to convert any distance x to a phase change by forming the product kx.

With an understanding of the meaning of the wave parameters and relationships described in this and the previous section, you're almost ready for a discussion of wavefunctions. But that discussion will be more meaningful to you if you also have a basic understanding of vector concepts, complex numbers, and the Euler relations. Those are the subjects of the next three sections.

1.3 Vector concepts

Before getting into complex numbers and Euler's relation, we think a discussion of basic vector concepts will provide a helpful foundation for those topics. That's because every complex number can be considered to be the result of vector addition, which is described later in this section. Furthermore, some waves involve vector quantities (such as electric and magnetic fields), and a quick review of the basics of vectors may help you understand those waves.

So what exactly is a vector? For many physics applications, you can think of a vector simply as a quantity that includes both a magnitude (how much) and a direction (which way). For example, speed is not a vector quantity; it's called a "scalar" quantity because it has magnitude (how fast an object is moving) but no direction. But velocity is a vector quantity, because velocity includes both speed and direction (how fast an object is moving and in which direction).

The are many other quantities that can be represented by vectors, including acceleration, force, linear momentum, angular momentum, electric fields, and magnetic fields. Vector quantities are often represented pictorially as arrows, in which the length of the arrow is proportional to the magnitude of the vector and the orientation of the arrow shows the direction of the vector. In text, vector quantities are usually indicated either using bold script (such as $\mathbf{A}$) or by putting an arrow over the variable name (such as $\vec{A}$).

Just as you can perform mathematical operations such as addition, subtraction, and multiplication with scalars, you can also do these operations with vectors. The two operations most relevant to using vectors to understand complex numbers are vector addition and multiplication of a vector by a scalar.

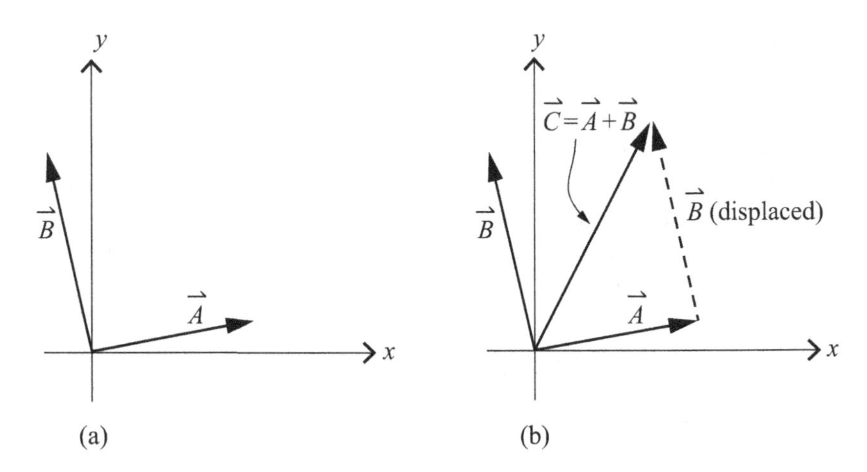

Figure 1.4 Graphical addition of vectors.

The simplest way to perform vector addition is to imagine moving one vector without changing its length or direction so that its tail (the end without the arrowhead) is at the head (the end with the arrowhead) of the other vector. The sum is then determined by making a new vector that begins at the tail of the first vector and terminates at the head of the second vector. This graphical "tail-to-head" approach to vector addition works for vectors in any direction and for three or more vectors as well.

To graphically add the two vectors $\vec{A}$ and $\vec{B}$ in Fig. 1.4(a), imagine moving vector $\vec{B}$ without changing its length or direction so that its tail is at the position of the head of vector $\vec{A}$, as shown in Fig. 1.4(b). The sum of these two vectors is called the "resultant" vector $\vec{C} = \vec{A} + \vec{B}$; note that $\vec{C}$ extends from the tail of $\vec{A}$ to the head of $\vec{B}$. The result would have been the same had you chosen to displace the tail of vector $\vec{A}$ to the head of vector $\vec{B}$ without changing the direction of $\vec{A}$.

It's extremely important for you to note that the length of the resultant vector is *not* the length of vector $\vec{A}$ added to the length of vector $\vec{B}$ (unless $\vec{A}$ and $\vec{B}$ happen to point in the same direction). So vector addition is not the same process as scalar addition, and you should remember to never add vectors using scalar addition.

Multiplication of a vector by a scalar is also quite straightforward, because multiplying a vector by any positive scalar does not change the direction of the vector – it only scales the length of the vector. Hence, $4\vec{A}$ is a vector in exactly the same direction as $\vec{A}$, but with length four times that of $\vec{A}$, as shown in Fig. 1.5(a). If the scaling factor is less than one the resulting vector is shorter

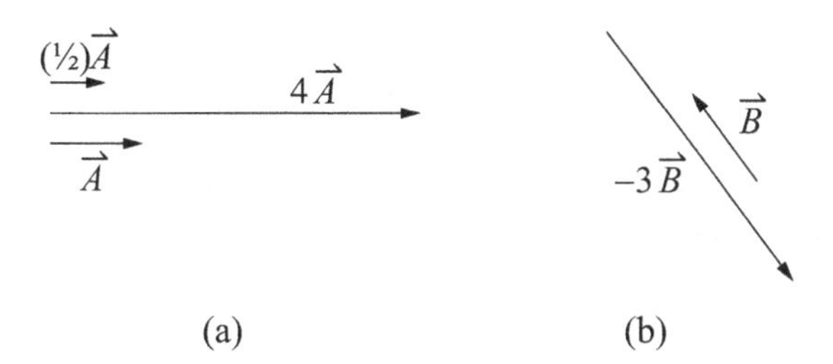

Figure 1.5 Multiplication of a vector by a scalar.

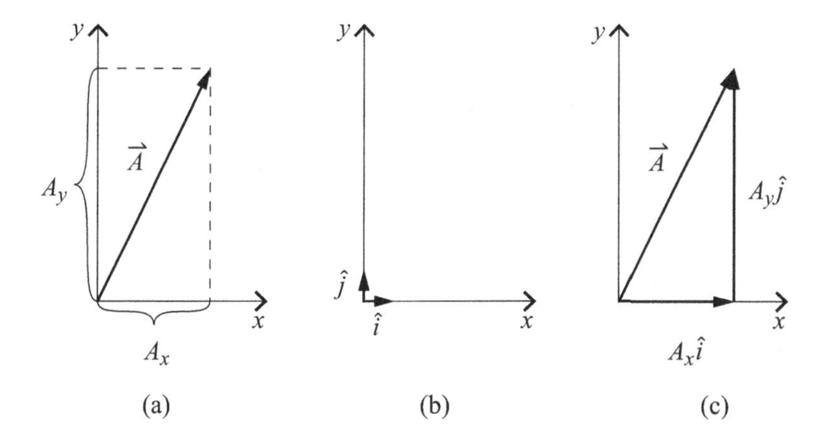

Figure 1.6 The vector $\vec{A}$ and its components A_x and A_y.

than the original vector, so multiplying $\vec{A}$ by (1/2) produces a vector that points in the same direction as $\vec{A}$ but is only half as long.

If the scalar multiplying factor is negative, the resulting vector is reversed in direction in addition to being scaled in length. So multiplying vector $\vec{B}$ by −3 produces the new vector $-3\vec{B}$, and that vector is three times as long as $\vec{B}$ and points in the opposite direction to that of $\vec{B}$, as shown in Fig. 1.5(b).

With vector addition and multiplication by a scalar in hand, you're in a position to understand an alternative way to express and manipulate vectors. That approach uses components and unit vectors and is directly relevant to the expression and manipulation of complex numbers.

To understand components and unit vectors in a two-dimensional Cartesian coordinate system, consider the vector $\vec{A}$ represented by the arrow shown in Fig. 1.6. As you can see in Fig. 1.6(a), the x-component (A_x) is the projection of vector $\vec{A}$ onto the x-axis (one way to see this is to imagine a light shining downward parallel to the y-axis and perpendicular to the x-axis, and to visualize the shadow cast by vector $\vec{A}$ onto the x-axis). Likewise, the y-component (A_y) is the projection of vector $\vec{A}$ onto the y-axis (in this case the light shines to the left, parallel to the x-axis and perpendicular to the y-axis).

Now look at Fig. 1.6(b). The two little arrows pointing along the x- and y-axes are called "basis vectors" or "unit vectors" because they have length of unity (that is, length of one unit). What are those units? They are whatever units the x- and y-axes have. So the $\hat{\imath}$ unit vector (called "i-hat") points along the x-axis and has length of one unit, and the $\hat{\jmath}$ unit vector (called "j-hat") points along the y-axis and has length of one unit. Be careful not to confuse $\hat{\imath}$ with i, which wears no hat and represents $\sqrt{-1}$.

The value of unit vectors becomes clear when you combine them with the vector components such as A_x and A_y. Since A_x and A_y are scalars, multiplying them by the unit vectors produces scaled versions of the unit vectors, as shown in Fig. 1.6(c). As you can see in that figure, $A_x\hat{\imath}$ is a vector in the direction of the unit vector $\hat{\imath}$ (that is, along the x-axis) but with length equal to A_x. Likewise, $A_y\hat{\jmath}$ is a vector in the direction of the unit vector $\hat{\jmath}$ (along the y-axis) but with length equal to A_y.

And here's the payoff: using vector addition, you can define the vector $\vec{A}$ as the vector sum of the vectors $A_x\hat{\imath}$ and $A_y\hat{\jmath}$. So a perfectly valid (and really useful) way to write vector $\vec{A}$ is

$$\vec{A} = A_x\hat{\imath} + A_y\hat{\jmath}. \tag{1.7}$$

In words, Eq. (1.7) says this: One way to get from the beginning to the end of vector $\vec{A}$ is to take A_x steps in the x-direction and then take A_y steps in the y-direction.

Once you have the x- and y-components of a vector, it's easy to find the magnitude (length) and direction (angle) of the vector. For the magnitude, just square the x- and y-components, add them, and take the square root (exactly as you do when using the Pythagorean theorem to find the length of the hypotenuse of a right triangle). The magnitude of a vector is written using vertical lines on each side of the vector's name (such as $|\vec{A}|$), so

$$|\vec{A}| = \sqrt{A_x^2 + A_y^2} \tag{1.8}$$

and the angle that $\vec{A}$ makes with the positive x-axis (measured anti-clockwise) is given by[3]

$$\theta = \arctan\left(\frac{A_y}{A_x}\right). \tag{1.9}$$

[3] You should be aware that most calculators have a "two-quadrant" arctan function, which means that you should add 180° to the output if the denominator (A_x in this case) is negative.

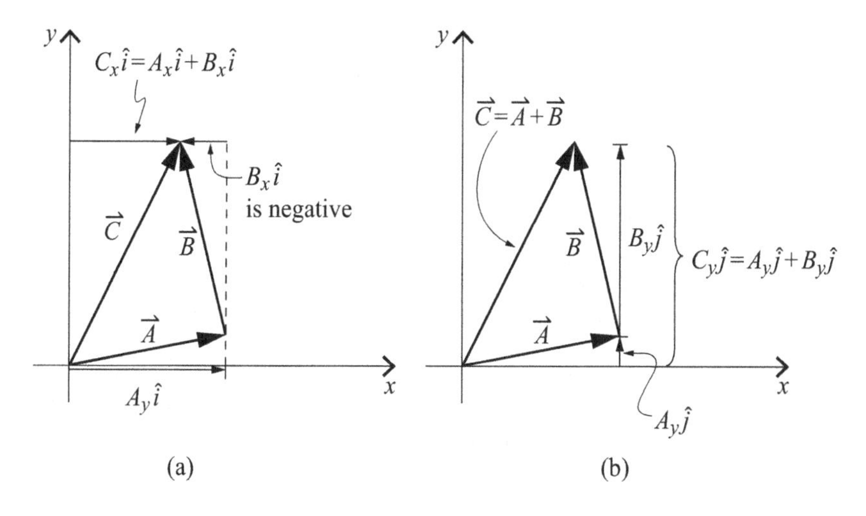

Figure 1.7 Component addition of vectors.

Expressing a vector using components and basis vectors makes the addition of vectors even easier than the graphical method described above. If vector $\vec{C}$ is the sum of two vectors $\vec{A}$ and $\vec{B}$, then the x-component of vector $\vec{C}$ (which is just C_x) is the sum of the x-components of vectors $\vec{A}$ and $\vec{B}$ (that is, $A_x + B_x$), and the y-component of vector $\vec{C}$ (called C_y) is the sum of the y-components of vectors $\vec{A}$ and $\vec{B}$ (that is, $A_y + B_y$). Thus

$$\begin{aligned} C_x &= A_x + B_x, \\ C_y &= A_y + B_y. \end{aligned} \tag{1.10}$$

The rationale for this is shown in Fig. 1.7.

Notice that it's possible to add the x-component of vector $\vec{A}$ to the x-component of vector $\vec{B}$ because these vector components both multiply $\hat{\imath}$ and the resulting vectors point in the same direction (along the x-axis). Likewise, you can add the y-component of vector $\vec{A}$ to the y-component of vector $\vec{B}$ because these vector components both multiply $\hat{\jmath}$ and the resulting vectors point in the same direction (along the y-axis). But you should never add the x-component of one vector to the y-component of that (or any other) vector.

Example 1.2 *If vector $\vec{F} = \hat{\imath} + 4\hat{\jmath}$ and vector $\vec{G} = -7\hat{\imath} - 2\hat{\jmath}$, what are the magnitude and direction of vector $\vec{H}$ that results from adding $\vec{F}$ to $\vec{G}$?*

Using the component approach, the x- and y-components of vector $\vec{H}$ are

$$\begin{aligned} H_x &= F_x + G_x = 1 - 7 = -6, \\ H_y &= F_y + G_y = 4 - 2 = 2, \end{aligned}$$

so $\vec{H} = -6\hat{\imath} + 2\hat{\jmath}$. Thus the magnitude of $\vec{H}$ is

$$|\vec{H}| = \sqrt{H_x^2 + H_y^2} = \sqrt{(-6)^2 + (2)^2} = 6.32$$

and the direction of $\vec{H}$ is

$$\theta = \arctan\left(\frac{H_y}{H_x}\right) = \arctan\left(\frac{2}{-6}\right) = -18.4^\circ,$$

but, since the denominator of the arctan argument is negative, the angle of vector $\vec{H}$ measured anti-clockwise from the positive x-axis is $-18.4^\circ + 180^\circ = 161.6^\circ$.

If you're wondering how vector components and vector addition are relevant to complex numbers, you can find the answer in the next section.

1.4 Complex numbers

An understanding of complex numbers can make the study of waves considerably less mysterious, and you probably already have an idea that complex numbers have real and imaginary parts. Unfortunately, the term "imaginary" often leads to confusion about the nature and usefulness of complex numbers. This section provides a review of complex numbers directed toward their use in wavefunctions and provides the foundation for the discussion of Euler's relation in the next section.

Many students are surprised to learn that the geometrical basis for complex numbers and the meaning of $\sqrt{-1}$ (typically called i in physics and mathematics and j in engineering) was first presented not by an ancient mathematician but by an eighteenth-century Norwegian–Danish surveyor and cartographer named Caspar Wessel.

In light of his occupation, it makes sense that Wessel spent a good deal of time thinking about the mathematics of directed line segments (the word "vector" was not yet in common use). Specifically, it was Wessel who developed the "head-to-tail" rule for vector addition described in Section 1.3. And, while imagining ways to multiply two directed line segments, Wessel hit upon a geometrical understanding of the meaning of $\sqrt{-1}$ that provides the basis for the concept of the complex plane.

To understand the complex plane, consider the number line shown in the left portion of Fig. 1.8. The locations of a few of the infinite number of real numbers are marked on this number line. Such number lines have been used for thousands of years, but Wessel (being a surveyor and mapmaker) thought

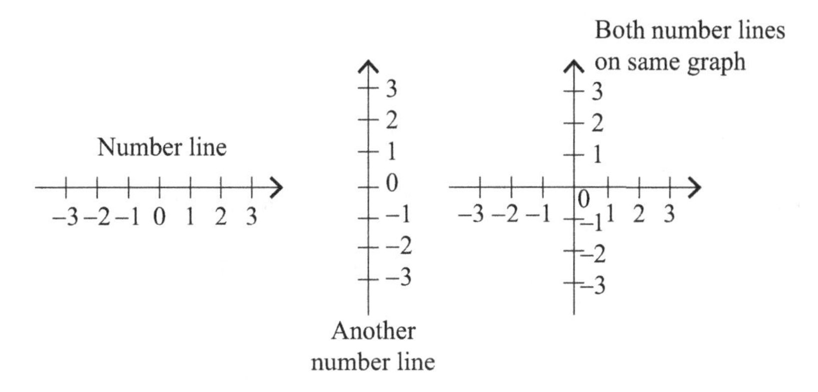

Figure 1.8 Number lines in two dimensions.

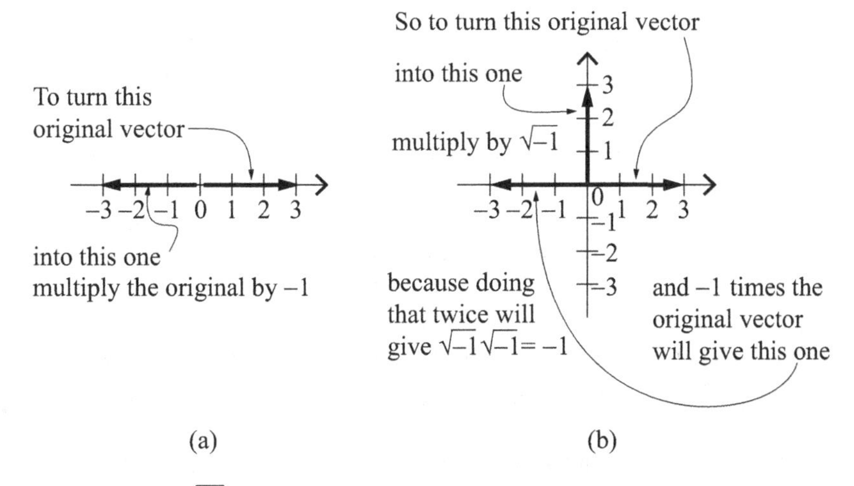

Figure 1.9 $\sqrt{-1}$ as a 90° rotation operator.

two-dimensionally. He imagined another number line at a 90° angle to the first, and then he imagined both number lines on the same graph, as shown in the right portion of Fig. 1.8.

Think about a directed line segment or vector such as the rightward-pointing arrow in Fig. 1.9(a). You know by the rules of scalar multiplication from Section 1.3 that, in order to reverse the direction of that original vector, you simply multiply it by –1, as shown in Fig. 1.9(a).

Now look at Fig. 1.9(b), which contains two perpendicular number lines, and imagine an operation that would rotate the original rightward-pointing vector by 90° so that it points along the vertical rather than the horizontal number line.

If you were then to perform the same operation on the (now-vertical) vector, it would again rotate by 90°, which means it would be pointing leftward along the horizontal number line.

But you know that reversing the direction of a vector is accomplished by multiplying that vector by –1. And, if a 180° rotation is accomplished by multiplying by –1, then each of the 90° rotations shown in the figure must correspond to a multiplication by $\sqrt{-1}$.

So a very useful way to think about i ($\sqrt{-1}$) is as an *operator* that produces a 90° rotation of any vector to which it is applied.

Thus the two perpendicular number lines form the basis of what we know today as the complex plane. Unfortunately, since multiplication by $\sqrt{-1}$ is needed to get from the horizontal to the vertical number line, the numbers along the vertical number line are called "imaginary". We say "unfortunately" because these numbers are every bit as real as the numbers along the horizontal number line. But the terminology is pervasive, so when you first learned about complex numbers, you probably learned that they consist of a "real" and an "imaginary" part.

This is usually written as

$$z = \mathrm{Re}(z) + i[\mathrm{Im}(z)], \tag{1.11}$$

in which the complex number z is seen to consist of a real part ($\mathrm{Re}(z)$) and an imaginary part ($\mathrm{Im}(z)$), with the i written explicitly to remind you that the imaginary portion is along a perpendicular number line.

If you compare Eq. (1.11) with Eq. (1.7) from Section 1.3, which was

$$\vec{A} = A_x\hat{\imath} + A_y\hat{\jmath}, \tag{1.7}$$

you may notice some similarity between these expressions. In both cases, the quantity on the left side of the equation (z or $\vec{A}$) is shown as equal to the sum of two terms on the right side. And, just as the two terms in the vector expression Eq. (1.7) refer to *different directions* and so cannot be added algebraically, so too the terms in the complex-number expression Eq. (1.11) refer to different number lines and cannot be added algebraically.

Happily, once you understand the mathematical operations defined for vectors, you can apply some of those same operations to complex numbers. The trick is to treat the real part of the complex number as the x-component of a vector and the imaginary part of the complex number as the y-component of a vector. So if you wish to find the magnitude of a complex number, you can use

$$|z| = \sqrt{[\mathrm{Re}(z)]^2 + [\mathrm{Im}(z)]^2} \tag{1.12}$$

and the angle of the complex number from the real axis is

$$\theta = \arctan\left(\frac{\text{Im}(z)}{\text{Re}(z)}\right). \tag{1.13}$$

An alternative method of finding the magnitude of a complex number involves the use of the number's "complex conjugate". The complex conjugate of a complex number is written with a superscript asterisk (such as z^*) and is obtained by reversing the sign of the imaginary portion of the number. Thus the complex number z and its complex conjugate z^* are

$$\begin{aligned} z &= \text{Re}(z) + i[\text{Im}(z)], \\ z^* &= \text{Re}(z) - i[\text{Im}(z)]. \end{aligned} \tag{1.14}$$

To find the magnitude of a complex number, simply multiply the complex number by its complex conjugate and take the square root of the result:

$$|z| = \sqrt{z^*z}. \tag{1.15}$$

To see that this approach is consistent with Eq. (1.12), multiply the real and imaginary components term-by-term:

$$\begin{aligned} |z| = \sqrt{z^*z} &= \sqrt{[\text{Re}(z) - i[\text{Im}(z)]][\text{Re}(z) + i[\text{Im}(z)]]} \\ &= \sqrt{[\text{Re}(z)][\text{Re}(z)] - i^2[\text{Im}(z)][\text{Im}(z)] + i[\text{Re}(z)][\text{Im}(z)] - i[\text{Re}(z)][\text{Im}(z)]} \\ &= \sqrt{[\text{Re}(z)][\text{Re}(z)] + [\text{Im}(z)][\text{Im}(z)]} = \sqrt{[\text{Re}(z)]^2 + [\text{Im}(z)]^2} \end{aligned}$$

in agreement with Eq. (1.12).

Since the real and imaginary parts of a complex number provide information relevant to two different number lines, graphing a complex number requires that both number lines be shown, as illustrated for several complex numbers in Fig. 1.10.

This is sometimes referred to as the rectangular form of complex-number graphing, but you may also convert the real and imaginary parts of a complex number into magnitude and angle using Eqs. (1.12) and (1.13), as shown in Fig. 1.11.

Alternatively, if you know the magnitude ($|z|$) and phase (θ) of a complex number z, the geometry of Fig. 1.11 illustrates that the real (Re) and imaginary (Im) parts of z may be found using

$$\begin{aligned} \text{Re}(z) &= |z|\cos\theta \\ \text{Im}(z) &= |z|\sin\theta. \end{aligned} \tag{1.16}$$

The polar form of a complex number is sometimes written as

$$\text{Complex Number} = \text{Magnitude}\angle\text{Angle}$$

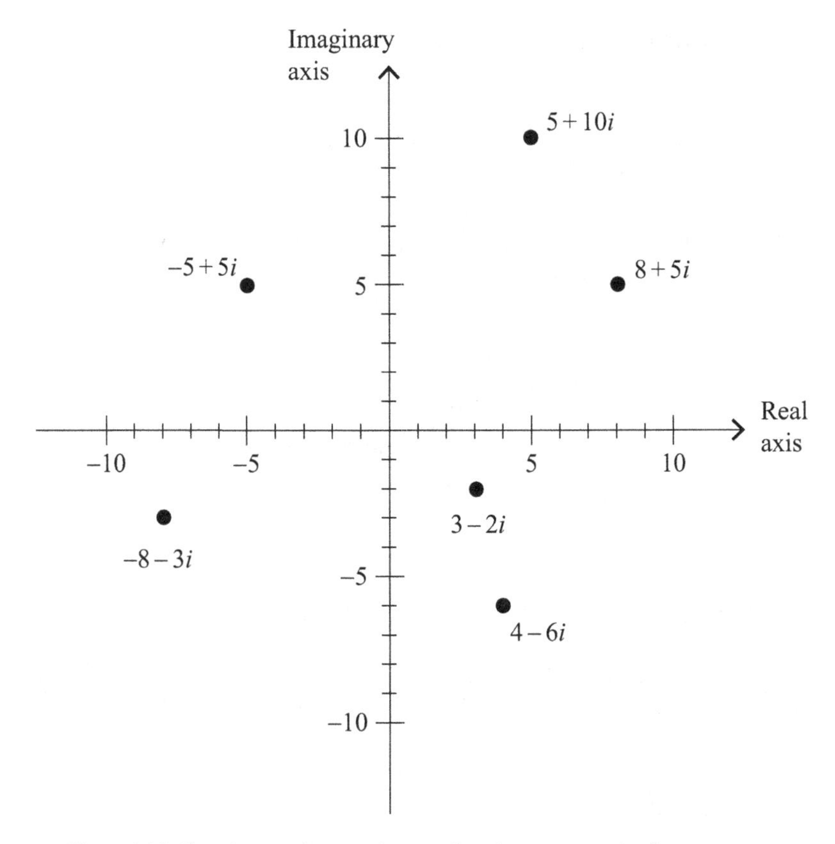

Figure 1.10 Complex numbers on the complex plane, rectangular form.

or

$$z = |z|\angle\theta. \tag{1.17}$$

Example 1.3 *Find the magnitude and angle of each of the complex numbers in Fig. 1.10.*

The rectangular-to-polar conversion equations (Eqs. (1.12) and (1.13)) can be applied to the complex numbers in Fig. 1.10 to determine the magnitude and angle of each. For the complex number $z = 5 + 10i$, $\text{Re}(z) = 5$ and $\text{Im}(z) = 10$, so for that number

$$|z| = \sqrt{[\text{Re}(z)]^2 + [\text{Im}(z)]^2} = \sqrt{(5)^2 + (10)^2} = 11.18$$

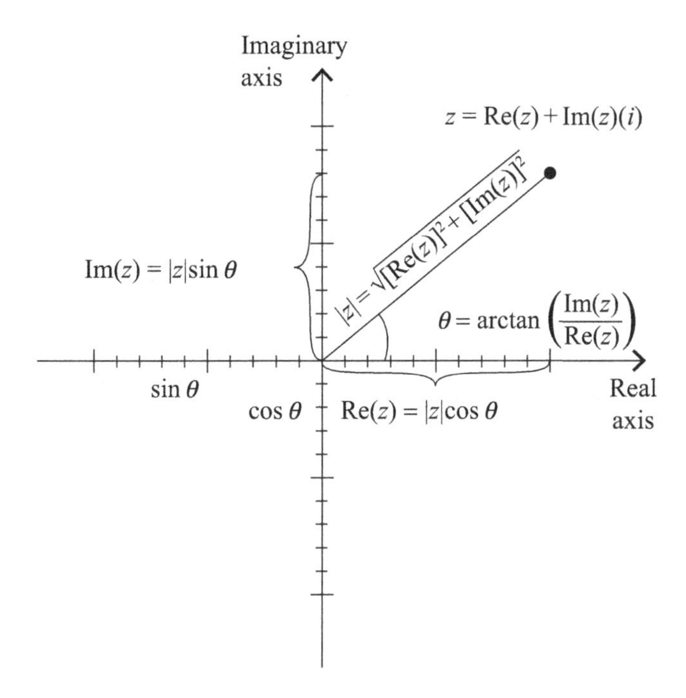

Figure 1.11 Converting from rectangular to polar form.

and the angle is

$$\theta = \arctan\left(\frac{\text{Im}(z)}{\text{Re}(z)}\right) = \arctan\left(\frac{10}{5}\right) = 63.4°.$$

Likewise, for the complex number $-5 + 5i$, $\text{Re}(z) = -5$ and $\text{Im}(z) = 5$, so

$$|z| = \sqrt{[\text{Re}(z)]^2 + [\text{Im}(z)]^2} = \sqrt{(-5)^2 + (5)^2} = 7.07$$

and the angle from the real axis is

$$\theta = \arctan\left(\frac{\text{Im}(z)}{\text{Re}(z)}\right) = \arctan\left(\frac{5}{-5}\right) = -45°.$$

Once again, since the denominator of the arctan argument is negative, the angle measured anti-clockwise from the positive real axis is $-45° + 180° = 135°$.

The magnitude and angle values for all six of the complex numbers in Fig. 1.10 are shown in Fig. 1.12.

With an understanding of the relationship between complex numbers and the complex plane, it's very useful to consider a special subset of the infinite

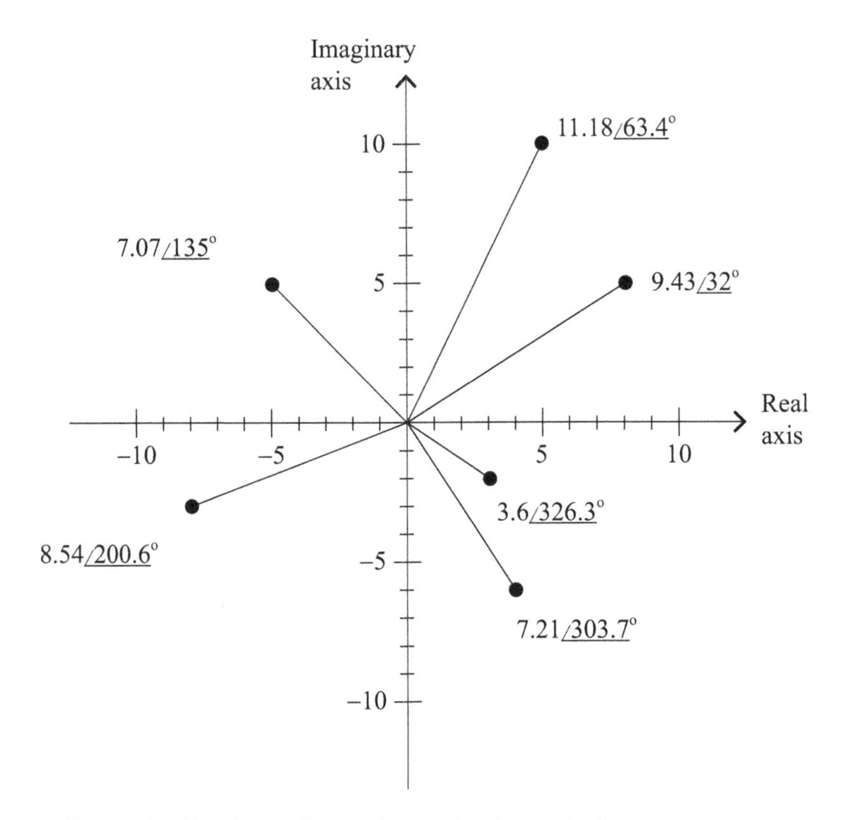

Figure 1.12 Complex numbers on the complex plane, polar form.

number of points in the complex plane. That subset is comprised of all the points that form a circle around the origin at a distance of exactly one unit. That circle of points is called the "unit circle" because its radius has unit length.

To see the usefulness of the unit circle, consider Fig. 1.13. Any complex number z lying on the unit circle can be drawn as a vector with length (magnitude) one and angle θ. Using Eqs. (1.16), the real and imaginary components of any number on the unit circle must be

$$\begin{aligned} \text{Re}(z) &= |z|\cos\theta = 1\cos\theta, \\ \text{Im}(z) &= |z|\sin\theta = 1\sin\theta, \end{aligned} \tag{1.18}$$

so any complex number on the unit circle can be written as

$$z = \cos\theta + i\sin\theta. \tag{1.19}$$

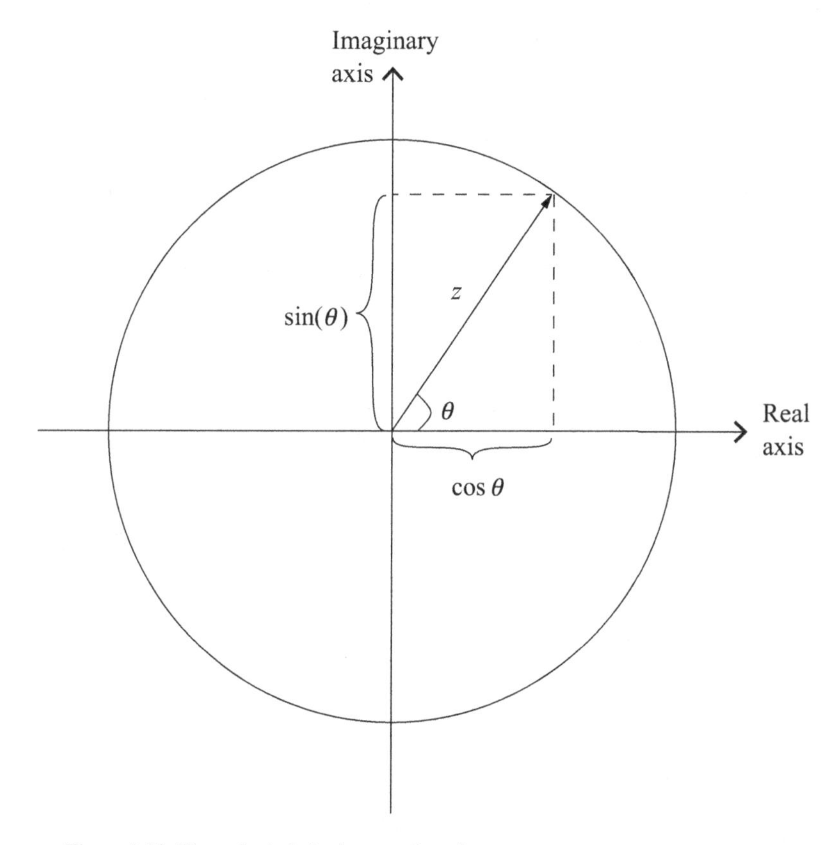

Figure 1.13 The unit circle in the complex plane.

If you aren't convinced that z has the correct magnitude, apply Eq. (1.15):

$$
\begin{aligned}
|z| &= \sqrt{z^*z} \\
&= \sqrt{(\cos\theta - i\sin\theta)(\cos\theta + i\sin\theta)} \\
&= \sqrt{\cos^2\theta + \sin^2\theta + i\sin\theta\cos\theta - i\sin\theta\cos\theta} \\
&= \sqrt{\cos^2\theta + \sin^2\theta} = \sqrt{1} = 1
\end{aligned}
$$

as expected for points on the unit circle.

The unit circle in the complex plane is especially useful in understanding a form of vectors called "phasors". Although different authors use different definitions for phasors, in most texts you'll find phasors described as vectors whose tips rotate around the unit circle in the complex plane. Such a phasor is shown in Fig. 1.14.

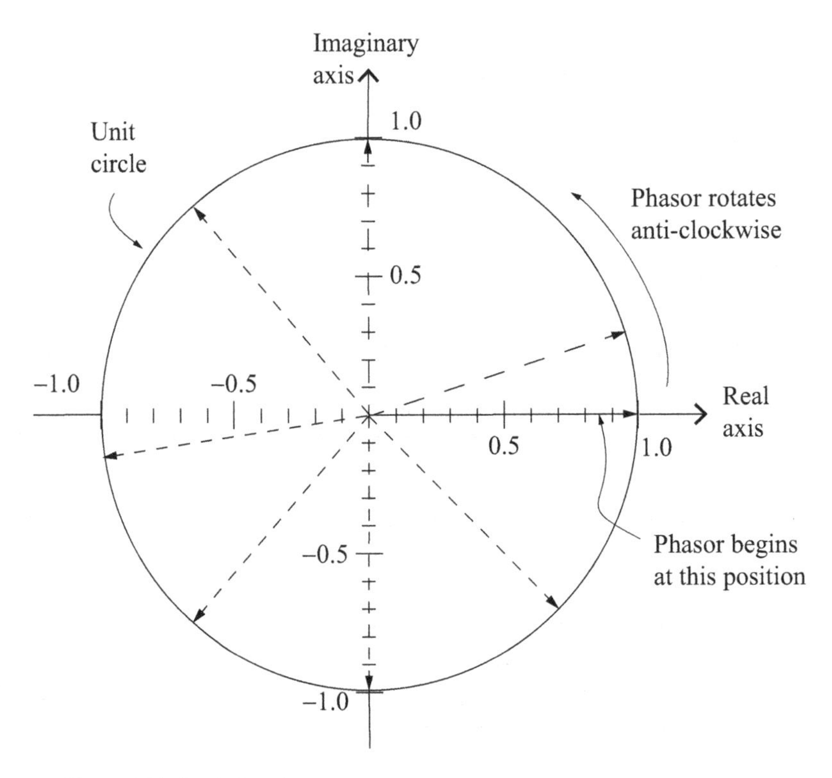

Figure 1.14 A rotating phasor.

Notice that in this figure the phasor extends from the origin to the unit circle; this particular phasor initially points along the positive real axis. As θ increases (becomes more positive), the phasor rotates anti-clockwise while maintaining its length of one unit. After one period, the phasor returns to the original position along the positive real axis.[4] If θ decreases (becomes less positive or more negative), the phasor rotates in the clockwise direction.

One reason why phasors are extremely useful in the analysis of waves is shown in Fig. 1.15. As shown in the right portion of the figure, as the phasor rotates, the projection of the phasor onto the imaginary axis traces out a sine wave as θ increases. And, as shown in the bottom portion of the same figure, the projection of the phasor onto the real axis traces out a cosine wave as θ advances.

[4] Depending on the application, other positions may be chosen as the original or "reference" position.

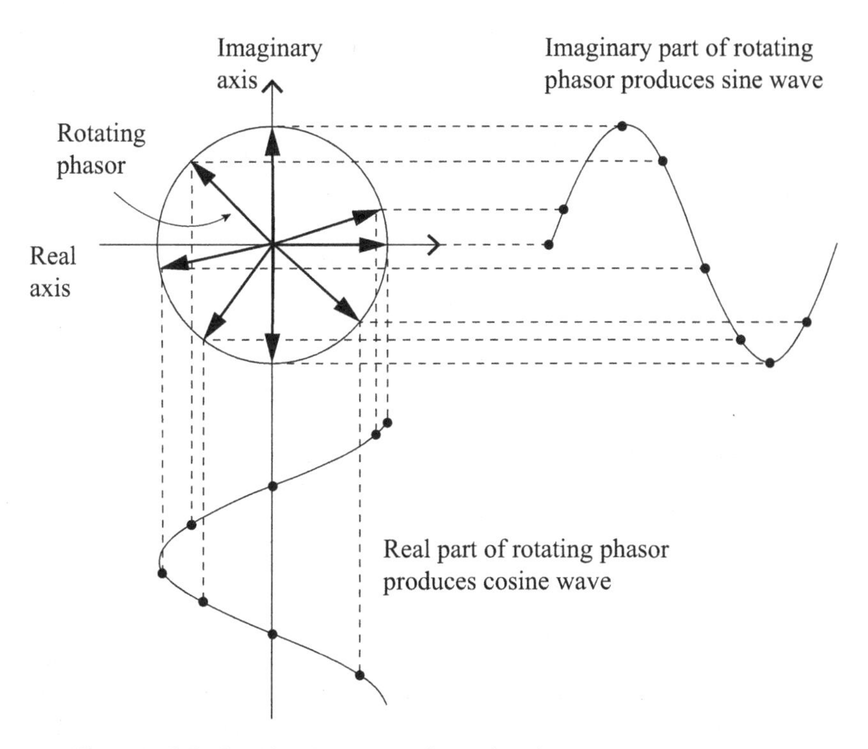

Figure 1.15 Real and imaginary parts of a rotating phasor.

Thus a rotating phasor can provide a useful representation of the advancing phase of a wave, and Euler's relation provides a means of performing mathematical operations using phasors. That's the subject of the next section.

1.5 Euler relations

While the notation used for z in the previous section is useful when you need separate real and imaginary components, it gets cumbersome very quickly when you start doing algebra or calculus involving z. It would be nice to package all the information you need to define z into an easier-to-use function of θ. Writing a complex number z as $z = \text{Mag}\angle\theta$ is a step in the right direction, but how do you perform mathematical operations (such as multiplying by another complex number or taking a derivative) on such an expression? For that, you'll need to express z as a function that has both magnitude and phase and is equivalent to the expression $z = \cos\theta + i\sin\theta$.

To find that function, one approach is to look at the behavior of its derivatives. The form of z involving sines and cosines has a first derivative of

$$\frac{dz}{d\theta} = -\sin\theta + i\cos\theta = i(\cos\theta + i\sin\theta) = iz \tag{1.20}$$

(remember that i times i is –1). The second derivative is

$$\frac{d^2z}{d\theta^2} = -\cos\theta - i\cos\theta = i^2(\cos\theta + i\sin\theta) = i^2z = -z, \tag{1.21}$$

so every time you take another derivative, you pick up a factor of i but otherwise leave the function unchanged.

At this point, to find z you could solve the differential equation $dz/d\theta = iz$ or you could guess a solution (a favorite method of solving differential equations among physicists). You can see how to solve the differential equation in one of the chapter-end problems and the solution on the book's website, but here's the rationale behind the guessing approach.

Equation (1.20) says that change in the function z as θ changes (that is, the slope $dz/d\theta$) is equal to the function (z) times a constant (i). Another factor of i is picked up each time a derivative with respect to θ is taken, which implies that i is multiplied by θ within the function. An initial guess might be that the function z is as simple as $i\theta$, but in that case θ would disappear after just one derivative. So a better guess for the function z might be something to the $i\theta$ power. This "something" – call it a – must be a very special value so that the derivative ($da^{i\theta}/d\theta$) brings down a factor of i but otherwise leaves the function unchanged.

Writing $i\theta$ as x, we're looking for a value of a for which $da^x/dx = a^x$. That means that the slope of the function must equal the value of the function for all values of x. A few choices for the value of a are shown in Fig. 1.16, which plots a^x versus x. If a is 2, then at $x = 1$ the value of a^x is $2^1 = 2$ and the slope, which is the derivative $d(2^x)/dx$ evaluated at $x = 1$, is 1.39 (too low). If a is 3, then at $x = 1$ the value of a^x is $3^1 = 3$ and the slope is 3.30 (too high). However, if a is in the sweet spot between 2 and 3, around 2.72, then the value of a^x at $x = 1$ is $2.72^1 = 2.72$ and the slope is 2.722. That's pretty close, and it turns out that, to get exactly the same value between the function and its slope, you have to use $a = 2.718$ plus an infinite amount of additional decimal places. Just as with π, it's easiest to give this irrational number a name: e, sometimes called "Euler's number".

Now the pieces are in place to construct a functional version of the complex number: $z = e^{i\theta}$. The first derivative of z with respect to θ is

$$\frac{dz}{d\theta} = i(e^{i\theta}) = iz \tag{1.22}$$

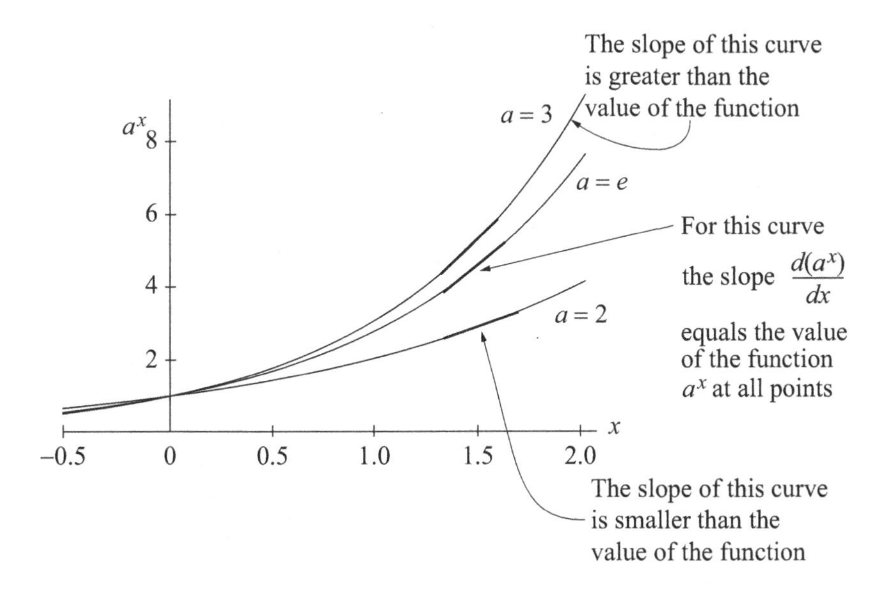

Figure 1.16 An illustration of why e is special.

and the second derivative is

$$\frac{d^2z}{d\theta^2} = i^2(e^{i\theta}) = i^2 z. \tag{1.23}$$

These are the same results as we obtained in Eqs. (1.20) and (1.21) by using $z = \cos\theta + i\sin\theta$. Setting these two versions of z equal to one another gives the Euler relation

$$e^{\pm i\theta} = \cos\theta \pm i\sin\theta. \tag{1.24}$$

This equation is considered by some mathematicians and physicists to be the most important equation ever devised. In Euler's relation, both sides of the equation are expressions for a complex number on the unit circle. The left side emphasizes the magnitude (the 1 multiplying $e^{i\theta}$) and direction in the complex plane (θ), while the right side emphasizes the real ($\cos\theta$) and imaginary ($\sin\theta$) components. Another approach to demonstrating the equivalence of the two sides of Euler's relation is to write out the power-series representation of each side; you can see how this works in the chapter-end problems and online solutions.

Before using the concepts developed in this and previous sections to discuss wavefunctions, we think it's worth a few minutes of your time to make sure you understand the importance of the presence of i in the exponent of the expression e^{ix}. Without it, the expression e^x is simply an exponentially

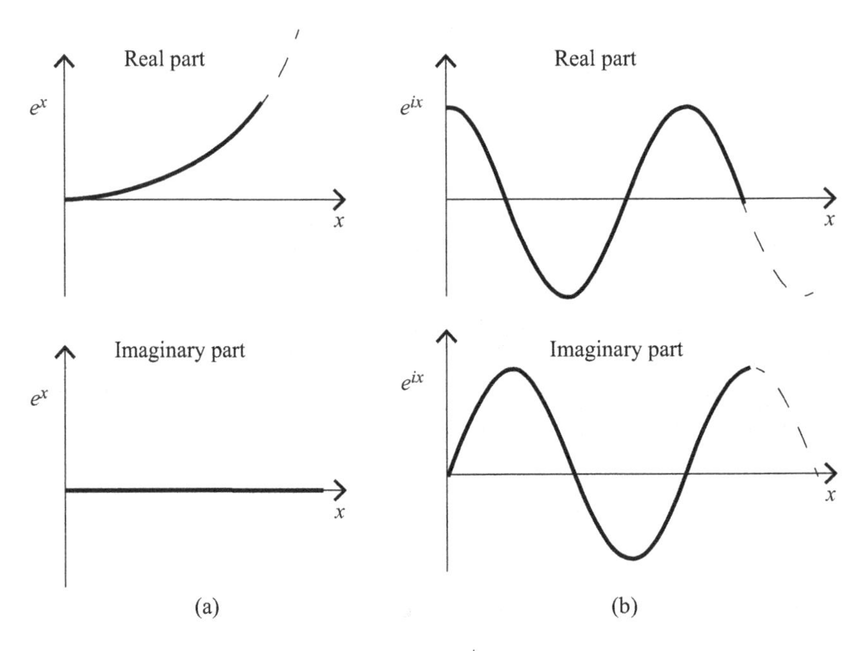

Figure 1.17 The difference between e^x and e^{ix}.

increasing real number as x increases, as shown in Fig. 1.17(a). But the presence of $\sqrt{-1}$ (the rotation operator between the two perpendicular number lines in the complex plane) in the exponent causes the expression e^{ix} to move from the real to the imaginary number line. As it does so, its real and imaginary parts oscillate in a sinusoidal fashion, as shown in Fig. 1.17(b). So the real and imaginary parts of the expression e^{ix} oscillate in exactly the same way as the real and imaginary components of the rotating phasor in Fig. 1.15.

1.6 Wavefunctions

The concept of the wavefunction is useful in many applications of wave theory, but students often express uncertainty about the nature and mathematical expression of wavefunctions. That uncertainty may arise from the terminology commonly used in wavefunction discussions, so the goal of this section is to provide both plain-language and mathematical definitions of wavefunctions and to illustrate their use in solving wave problems.

Stated simply, the wavefunction of any wave is the function that defines the value of the wave's disturbance at every place and time. When reading about wavefunctions, you'll often run into seemingly redundant statements such as

$$y(x, t) = f(x, t) \tag{1.25}$$

or

$$y = f(x, t), \tag{1.26}$$

or perhaps

$$\psi = f(x, t). \tag{1.27}$$

In these equations, y and ψ represent the displacement of the wave[5] and f does not represent frequency. Instead, f represents a function of position (x) and time (t). And exactly what is that function? It's whatever function describes the *shape* of the wave in time and space.

To understand that, just remember that saying "a function of x and t" just means "depends on x and t". So an equation such as $y = f(x, t)$ means that the value of the displacement (y) depends on both the location (x) and the time (t) at which you measure the wave.[6] So if the function f changes very slowly with x and with t, you'd have to look at the wave at two far-apart locations or at two very different times to see much difference in the disturbance produced by the wave.

And, since the choice of the function f determines the shape of the wave, Eqs. (1.25) through (1.27) tell you that the displacement of the wave at any place or time depends on the shape of the wave and its position.

The easiest way to think about the shape of a wave is to imagine taking a snapshot of the wave at some instant of time. To keep the notation simple, you can call the time at which the snapshot is taken $t = 0$; snapshots taken later will be timed relative to this first one. At the time of that first snapshot Eq. (1.26) can be written as

$$y = f(x, 0). \tag{1.28}$$

Many waves maintain the same shape over time – the wave moves in the direction of propagation, but all peaks and troughs move in unison, so the shape does not change as the wave moves. For such "non-dispersive" waves, $f(x, 0)$ can be written as $f(x)$, since the shape of the wave doesn't depend on when

[5] Remember that "displacement" doesn't necessarily refer to a physical dislocation; it refers to whatever disturbance is produced by the wave.

[6] In this section, we'll consider only waves moving along the x-axis, so any location may be specified by the value of x. This can be generalized to waves traveling in other directions.

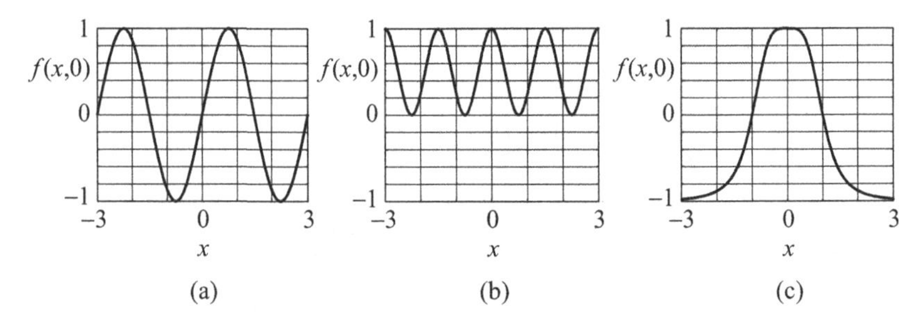

Figure 1.18 Graphs of the wave profiles of Eq. (1.29).

you take the snapshot. The function $f(x)$ can then be called the "wave profile". Some example wave profiles are

$$
\begin{aligned}
y &= f(x, 0) = A\sin(kx), \\
y &= f(x, 0) = A[\cos(kx)]^2, \\
y &= f(x, 0) = \frac{1}{ax^4 + b}.
\end{aligned}
\tag{1.29}
$$

Graphs of these wave profiles are shown in Fig. 1.18. These wave profiles look quite similar to the previous figures of waves, such as Fig. 1.1. That's a limitation of showing functions of space and time in a static environment, such as this book. To see the difference between a wave and a wave profile, imagine drawing a few periods of a sine wave, as in Fig. 1.18(a). If you repeat the drawing on separate pieces of paper with the wave profile shifted to the left or right by a small amount each time, you can put the pieces together to create a "flipbook". Flipping through the book shows the wave itself as it moves over time, while each page shows the wave profile at a certain instant in time.

Once you have a wave profile $f(x)$, it's a short step to the wavefunction, $y(x, t)$. To make that step, you have to think about the way to include the fact that the disturbance produced by the wave depends both on space and on time. The definitions in the first section of this chapter provide a hint about how to do that. Recall that displacement answers the question "how big is the wave at this place and time?" and phase answers the question "what part of the wave (e.g. crest or trough) is at this place and time?" So it makes sense that the functional dependence of the displacement is related to the phase of the wave.

To make this explicit, recall from Section 1.2 that the phase of a wave depends on space ($\Delta\phi_{\text{const } t} = k\,\Delta x$) and on time ($\Delta\phi_{\text{const } x} = \omega\,\Delta t$). Thus the total phase change over both space and time can be written as

$$\Delta\phi = \phi - \phi_0 = k\,\Delta x \pm \omega\,\Delta t, \tag{1.30}$$

where the "$\pm$" allows for wave propagation in either direction, as discussed later in this section. Writing the change in location as $\Delta x = x - x_0$ and the change in time as $\Delta t = t - t_0$, you can choose to set $x_0 = 0$ and $t_0 = 0$, which makes $\Delta x = x$ and $\Delta t = t$. If the starting phase ϕ_0 is taken as zero, the phase at any location (x) and time (t) can be written as

$$\phi = kx \pm \omega t. \tag{1.31}$$

Thus the functional form of the displacement can be written as

$$y(x, t) = f(kx \pm \omega t), \tag{1.32}$$

where the function f determines the shape of the wave and the argument of the function (that is, $kx \pm \omega t$) is the phase ϕ of the wave at each location (x) and time (t).

This equation is extremely useful in solving a wide range of wave problems, and it has the speed of the wave built right in. Before describing how to find the wave speed and direction in this expression, here's an example of a specific wavefunction and how to use it to determine the wave disturbance at a given place and time.

Example 1.4 *Consider a wave with wavefunction given by*

$$y(x, t) = A\sin(kx + \omega t), \tag{1.33}$$

where the wave amplitude (A) is 3 meters, the wavelength (λ) is 1 meter, and the wave period (T) is 5 seconds. Find the value of the displacement $y(x, t)$ at the position $x = 0.6$ m and time $t = 3$ seconds.

One approach would be make a flipbook of this wave. The wave amplitude tells you how big to make the peaks of your wave, the wavelength tells you how far apart to space the peaks on each page, and the wave period tells you how much to shift the wave between the pages of your book (since it has to move a distance of one wavelength in the direction of propagation during each period). You could then turn to the page in your flipbook corresponding to a time of 3 seconds and measure the y-value (the displacement) of the wave at a distance of 0.6 meters along the x-axis.

Alternatively, you can just plug each of the variables into Eq. (1.33). The wavelength of 1 meter means that the wavenumber is $k = 2\pi/1 = 2\pi$ rad/m, and the wave period of 5 seconds tells you that the frequency is $f = 1/5 = 0.2$ Hz (and the angular frequency is $\omega = 2\pi f = 0.4\pi$ rad/s). Plugging in these values gives

$$\begin{aligned} y(x,t) &= A\sin(kx+\omega t) \\ &= (3\text{ m})\sin[(2\pi\text{ rad/m})(0.6\text{ m}) + (0.4\pi\text{ rad/s})(3\text{ s})] \\ &= (3\text{ m})\sin(2.4\pi\text{ rad}) = 2.85\text{ m}. \end{aligned}$$

To understand how the wave speed and direction are built into Eq. (1.32), it helps to understand what happens to a function such as $f(x)$ when you add or subtract another term in the argument, such as $f(x+1)$ or $f(x-1)$. Take a look at the table of values and graph of the triangular-pulse function shown in Fig. 1.19.

Now imagine what happens if you make a similar table and graph for the function $f(x+1)$. Many students assume that this will shift the function toward the right (that is, in the positive x-direction), since we're somehow "adding 1 to x". But, as you can see in Fig. 1.20, the opposite is true. Adding a constant with the same sign as the x-term (in this case, there's a "+" sign in front of both the x-term and the constant (1) is positive) inside the argument shifts the function toward the left (that is, in the negative x-direction).

You can see why that's true by looking at the values in the table in Fig. 1.20. Since you're adding +1 to x inside the argument of the function, the function reaches a given value at a *smaller* value of x. Hence it shifts to the left, not to the right.

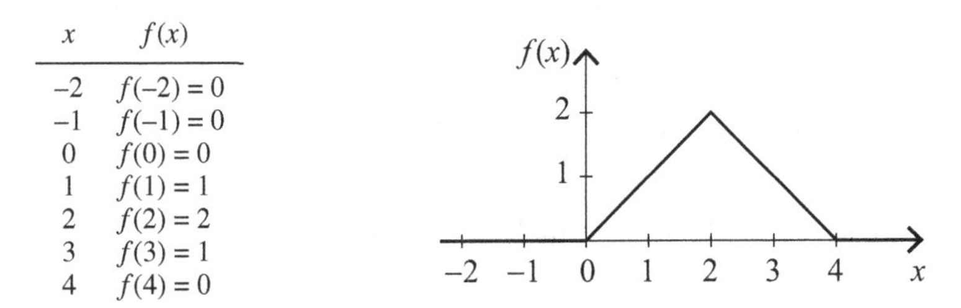

x	$f(x)$
–2	$f(-2)=0$
–1	$f(-1)=0$
0	$f(0)=0$
1	$f(1)=1$
2	$f(2)=2$
3	$f(3)=1$
4	$f(4)=0$

Figure 1.19 $f(x)$ vs. x for a triangular-pulse waveform.

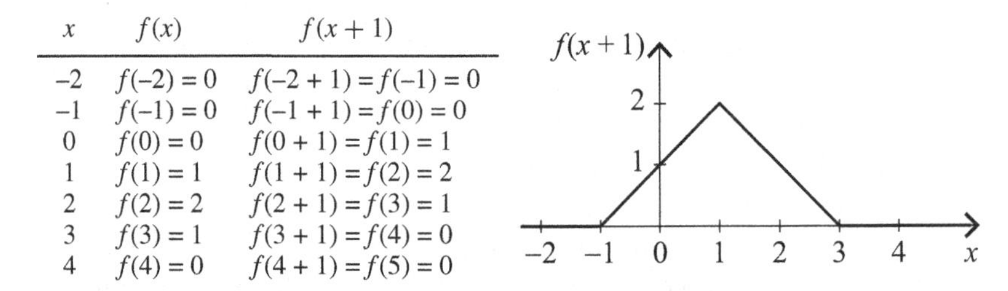

x	$f(x)$	$f(x+1)$
–2	$f(-2)=0$	$f(-2+1)=f(-1)=0$
–1	$f(-1)=0$	$f(-1+1)=f(0)=0$
0	$f(0)=0$	$f(0+1)=f(1)=1$
1	$f(1)=1$	$f(1+1)=f(2)=2$
2	$f(2)=2$	$f(2+1)=f(3)=1$
3	$f(3)=1$	$f(3+1)=f(4)=0$
4	$f(4)=0$	$f(4+1)=f(5)=0$

Figure 1.20 $f(x+1)$ vs. x for a triangular-pulse waveform.

x	$f(x)$	$f(x-1)$
−2	$f(-2)=0$	$f(-2-1)=f(-3)=0$
−1	$f(-1)=0$	$f(-1-1)=f(-2)=0$
0	$f(0)=0$	$f(0-1)=f(-1)=0$
1	$f(1)=1$	$f(1-1)=f(0)=0$
2	$f(2)=2$	$f(2-1)=f(1)=1$
3	$f(3)=1$	$f(3-1)=f(2)=2$
4	$f(4)=0$	$f(4-1)=f(3)=1$

Figure 1.21 $f(x-1)$ vs. x for a triangular-pulse waveform.

Following that same logic, you should be able to see why the function $f(x-1)$ is shifted to the right (in the positive x-direction) as shown in Fig. 1.21.

So does adding a positive constant always shift the function in the negative x-direction? Not if you have a function such as $f(-x+1)$. In that case, the function shifts in the positive x-direction relative to $f(-x)$ because the sign of the x-term is negative and the constant (1) is positive. Likewise, $f(-x-1)$ shifts in the negative x-direction (if you don't see why that's true, take a look at the chapter-end problems and online solutions).

This means you can't determine whether a function is shifted left or right just by looking at the sign of the additional term in the argument; you have to compare the sign of that term with the sign of the x-term. If those signs are the same, the function is shifted in the negative x-direction, and if those signs are opposite, the function is shifted in the positive x-direction.

With this in mind, you should see that a function such as

$$y(x,t) = f(kx + \omega t) \tag{1.34}$$

represents a wave moving in the negative x-direction (since the signs on the x-term and the time-term are the same), while

$$y(x,t) = f(kx - \omega t) \tag{1.35}$$

represents a wave moving in the positive x-direction (since the signs on the x-term and the time-term are opposite).[7]

Knowing the wave direction is helpful, but there's more information in Eq. (1.32). Specifically, the speed (v) with which a given point on the wave moves (called the "phase speed" of the wave, as described in Section 1.1) can be found directly from this equation.

To dig the wave phase speed (v) out of Eq. (1.32), it helps to recall that the definition of speed is the distance covered divided by the amount of time it

[7] Remember, the values of the variables x and t may be positive or negative, but it's the signs in front of these terms that matter.

takes to cover that distance. You know both of those quantities (distance and time) for a single cycle of a wave, because the wave covers a distance of one wavelength (λ) in an amount of time equal to one wave period (T). Dividing the distance by the time gives $v = \lambda/T$. From Section 1.2 you know that $\lambda = 2\pi/k$ and $T = 2\pi/\omega$. Putting these three expressions together gives

$$v = \frac{\lambda}{T} = \frac{2\pi/k}{2\pi/\omega} = \frac{2\pi}{k}\frac{\omega}{2\pi}$$

or

$$v = \frac{\omega}{k}. \tag{1.36}$$

This is a very useful result. If you know both the angular frequency (ω) and the wavenumber (k), you can divide ω by k to find the wave phase speed. And, if you're given a function such as that shown in Eq. (1.32) and you need the wave speed, just take whatever multiplies the time term (t) and divide it by whatever multiplies the position term (x).

As always, it's useful to ask yourself whether the units work. Rewriting Eq. (1.36) in terms of SI units gives

$$\left[\frac{\text{meters}}{\text{second}}\right] = \left[\frac{\cancel{\text{radians}}}{\text{second}} \times \frac{\text{meters}}{\cancel{\text{radian}}}\right], \tag{1.37}$$

leaving m/s on both sides.

You may also encounter a version of Eq. (1.32) in which the wave phase speed appears explicitly. To see how that works, remember that the angular frequency (ω) equals $2\pi f$, and $f = v/\lambda$. Thus $\omega = 2\pi v/\lambda = kv$, so Eq. (1.32) may be written as

$$\begin{aligned} y(x,t) &= f(kx \pm \omega t) \\ &= f(kx \pm kvt) = f[k(x \pm vt)], \end{aligned}$$

which is sometimes written as

$$y(x,t) = f(x \pm vt). \tag{1.38}$$

You may have noticed that in Eq. (1.38) the argument of f is no longer a phase, since x and vt both have units of distance, not radians. That's only because the purpose of this equation is to show the functional dependence of the displacement (f) on x and t, so the factor of the wavenumber (k) is not explicitly shown. But you can always convert the distances in the argument ($x \pm vt$) to phase (with units of radians) by multiplying by the wavenumber (k).

1.7 Phasor representation of wavefunctions

Putting together the concepts of the complex plane, the Euler relations, and phasors provides a very powerful tool for the analysis of wavefunctions. To understand how this works, consider two waves represented by the following wavefunctions:

$$\begin{aligned} y_1(x,t) &= A_1 \sin(k_1 x + \omega_1 t + \epsilon_1), \\ y_2(x,t) &= A_2 \sin(k_2 x + \omega_2 t + \epsilon_2). \end{aligned} \tag{1.39}$$

If the amplitudes of these waves are equal (so $A_1 = A_2 = A$), and if the waves also have the same wavelength (hence the same wavenumber, so $k_1 = k_2 = k$) and the same frequency (hence the the same angular frequency, so $\omega_1 = \omega_2 = \omega$), then the only difference between the waves must be due to their phase constants (ϵ_1 and ϵ_2). Taking ϵ_1 as zero and ϵ_2 as $\pi/2$, the wavefunctions are

$$\begin{aligned} y_1(x,t) &= A \sin(kx + \omega t), \\ y_2(x,t) &= A \sin(kx + \omega t + \pi/2). \end{aligned} \tag{1.40}$$

To plot such wavefunctions on a two-dimensional graph, you have to decide whether you wish to see how the wavefunctions behave as a function of distance (x) or as a function of time (t). If you choose to plot them versus distance, you must choose the value of time at which you wish to see the functions. At time $t = 0$, the wavefunctions are

$$\begin{aligned} y_1(x,0) &= A \sin(kx), \\ y_2(x,0) &= A \sin(kx + \pi/2) \end{aligned} \tag{1.41}$$

and have the shapes shown in Fig. 1.22.

Notice that the phase constant ($\epsilon = \pi/2$) has the effect of shifting the second wavefunction (y_2) to the left, since the phase constant and the x-term have the same sign (that is, they are both positive). Notice also that a positive phase shift of $\pi/2$ radians (90°) has the effect of turning a sine function into a cosine function, since $\cos(\theta) = \sin(\theta + \pi/2)$.

Now consider what happens when you choose to plot these waves as a function of time (t) rather than distance (x). Just as you had to pick a specific time when plotting versus distance, now you must select a specific value of x at which your time plot will apply. Selecting $x = 0$, the wavefunctions are

$$\begin{aligned} y_1(0,t) &= A \sin(\omega t), \\ y_2(0,t) &= A \sin(\omega t + \pi/2). \end{aligned} \tag{1.42}$$

These functions are plotted individually in Fig. 1.23, but it's easier to compare waves by plotting them on the same graph, as in Fig. 1.24. In this figure, the

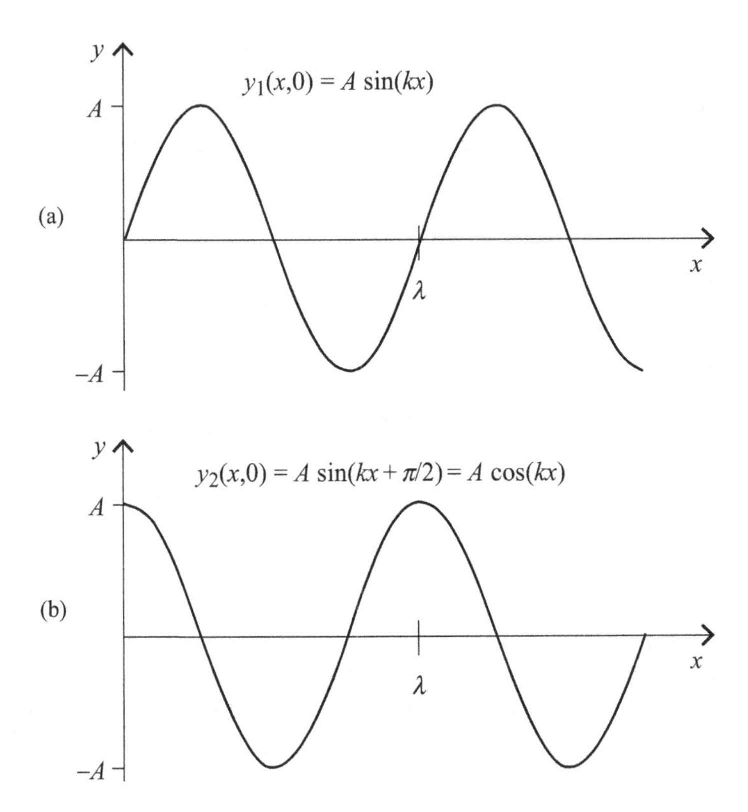

Figure 1.22 Waveforms of Eq. (1.40) at time $t = 0$.

first (y_1) wavefunction has been plotted using a dashed line to distinguish it from the second (y_2) wavefunction. When comparing waveforms on a time-domain plot such as Fig. 1.24, you may encounter terminology such as "y_2 is leading y_1" or "y_1 is lagging y_2". Many students find this confusing, since in this plot it appears that y_1 is somehow "ahead" of y_2 (that is, the peaks of y_1 appear to the right of the peaks of y_2). To see the flaw is this logic, remember that time is increasing to the right in this plot, so the peaks of y_2 occur *before* (that is, to the left of) the peaks of y_1. Hence in a plot with time increasing to the right, "leading" waves appear to the left of "lagging" waves.[8]

Phasor diagrams can be extremely helpful when analyzing wavefunctions such as y_1 and y_2. But, since these wavefunctions are written as sine or cosine waves and the phasors discussed in Section 1.4 were shown to represent e^{ix},

[8] If you're thinking that there are other peaks of y_2 to the right of the peaks of y_1, remember that you should always compare the positions of the *closest* peaks.

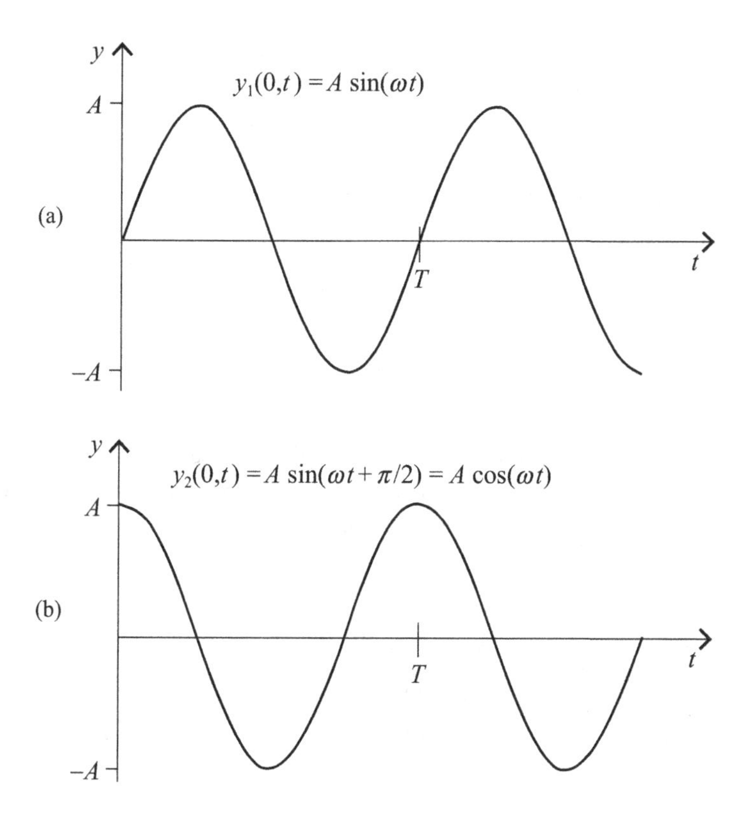

Figure 1.23 The waveforms of Eq. (1.40) at location $x = 0$.

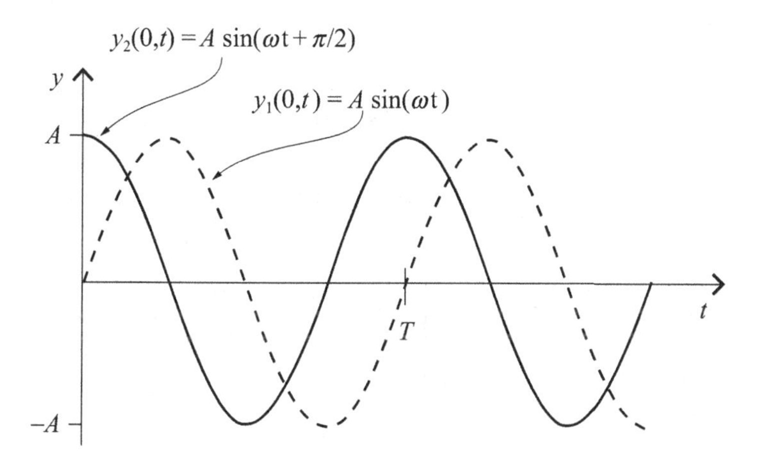

Figure 1.24 Leading and lagging in time.

it may not be clear how to represent y_1 and y_2 as phasors. To understand that, recall the Euler relation

$$e^{\pm i\theta} = \cos\theta \pm i\sin\theta \tag{1.24}$$

and observe what happens if you add $e^{i\theta}$ to $e^{-i\theta}$:

$$\begin{aligned} e^{i\theta} + e^{-i\theta} &= (\cos\theta + i\sin\theta) + (\cos\theta - i\sin\theta) \\ &= \cos\theta + \cos\theta + i\sin\theta - i\sin\theta = 2\cos\theta \end{aligned}$$

or

$$\cos\theta = \frac{e^{i\theta} + e^{-i\theta}}{2}. \tag{1.43}$$

This is useful. It shows that the cosine function can be represented by two counter-rotating phasors, since as θ increases $e^{i\theta}$ rotates anti-clockwise and $e^{-i\theta}$ rotates clockwise, as illustrated in Fig. 1.25.

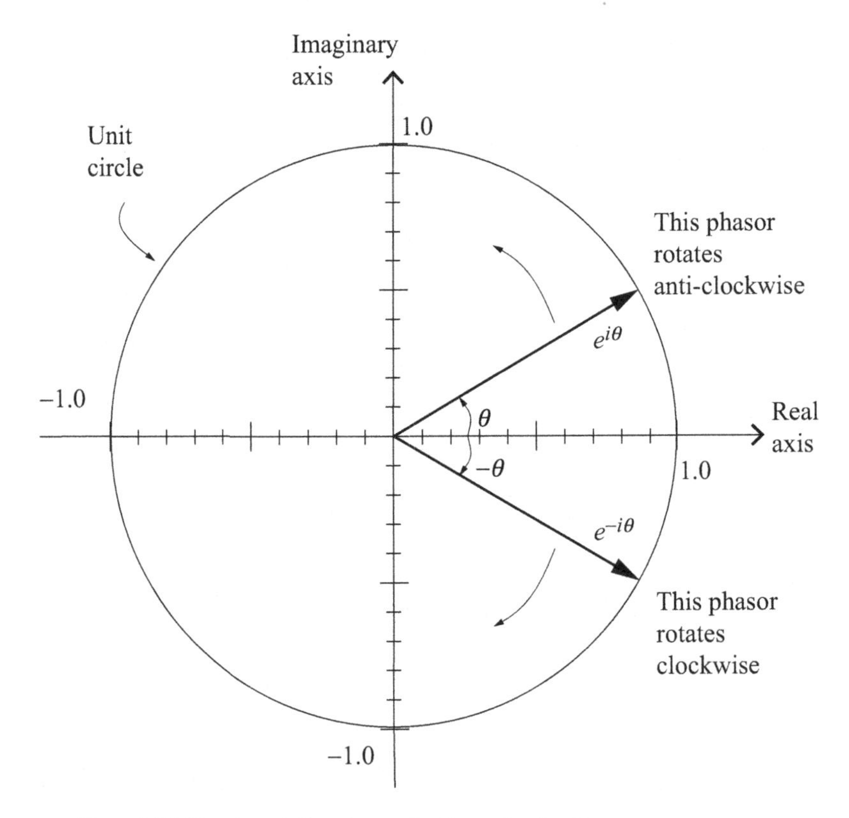

Figure 1.25 Counter-rotating phasors for the cosine function.

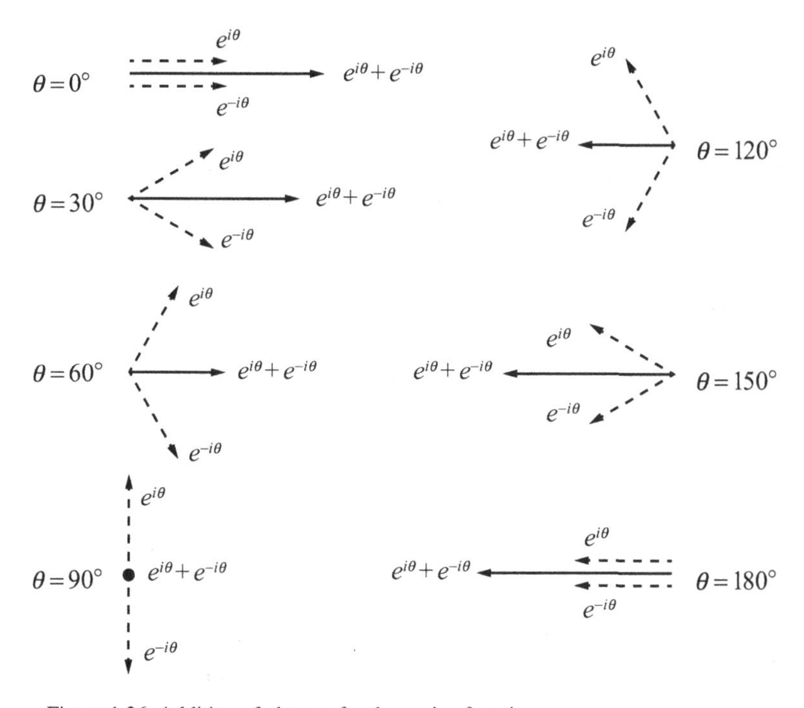

Figure 1.26 Addition of phasors for the cosine function.

To understand how the counter-rotating phasors $e^{i\theta}$ and $e^{-i\theta}$ add together for various values of θ, take a look at Fig. 1.26. At $\theta = 0$, both of these phasors point along the positive horizontal (real) axis of the complex plane (for clarity the phasors $e^{i\theta}$ and $e^{-i\theta}$ are drawn using dashed lines and slightly offset from one another in Fig. 1.26). Adding these two same-direction phasors produces a resultant phasor (which we label $e^{i\theta} + e^{-i\theta}$) that also points along the positive real axis and has a magnitude (length) of 2 (since $e^{i\theta}$ and $e^{-i\theta}$ both have a magnitude of 1). Hence the expression $(e^{i\theta} + e^{-i\theta})/2$ has a magnitude of 1, which is the value of $\cos\theta$ for $\theta = 0$.

Now look at the phasors $e^{i\theta}$ and $e^{-i\theta}$ for $\theta = 30°$. The phasor $e^{i\theta}$ points above the real axis by 30°, and the phasor $e^{-i\theta}$ points below the real axis by 30°. So the positive vertical (imaginary) component of $e^{i\theta}$ cancels out the negative vertical component of $e^{-i\theta}$, while the positive horizontal (real) component of $e^{i\theta}$ adds to the positive horizontal component of $e^{-i\theta}$ to give a resultant that points along the positive real axis with magnitude of 1.73. Dividing this value by 2 gives 0.866, which is the value of $\cos\theta$ for $\theta = 30°$.

Increase θ to 60° and the vertical (imaginary) components of $e^{i\theta}$ and $e^{-i\theta}$ still cancel out, while the horizontal (real) components add to the smaller value of 1.0, and dividing by 2 gives 0.5, the value of $\cos\theta$ for $\theta = 60°$.

For $\theta = 90°$, the horizontal (real) components are zero both for $e^{i\theta}$ and for $e^{-i\theta}$, and the vertical (imaginary) components still cancel out, so the value of the resultant $e^{i\theta} + e^{-i\theta}$ is zero, which is the value of $\cos\theta$ for $\theta = 90°$.

The same analysis applies for θ between 90° and 180°; for those angles the resultant phasor $e^{i\theta} + e^{-i\theta}$ points along the negative real axis, as you can see on the right side of Fig. 1.26. As θ increases from 180° to 360°, the $e^{i\theta}$ and $e^{-i\theta}$ phasors continue rotating, and the resultant $e^{i\theta} + e^{-i\theta}$ returns to zero and then to 1, as expected for the cosine function at these angles.

So the cosine function can be represented by the two counter-rotating phasors $e^{i\theta}$ and $e^{-i\theta}$, and at any angle the addition of those phasors and division by 2 yields the value of the cosine of that angle. Can the sine function also be represented in this way?

Yes it can. To see how that works, observe what happens if you subtract $e^{-i\theta}$ from $e^{i\theta}$:

$$\begin{aligned} e^{i\theta} - e^{-i\theta} &= (\cos\theta + i\sin\theta) - (\cos\theta - i\sin\theta) \\ &= \cos\theta - \cos\theta + i\sin\theta - (-i\sin\theta) = 2i\sin\theta \end{aligned}$$

or

$$\sin\theta = \frac{e^{i\theta} - e^{-i\theta}}{2i}. \tag{1.44}$$

This result shows that the sine function can also be represented by two counter-rotating phasors, since as θ increases $e^{i\theta}$ rotates anti-clockwise and $-e^{-i\theta}$ rotates clockwise, as illustrated in Fig. 1.27. Notice that in this figure, in addition to the phasors $e^{i\theta}$ and $e^{-i\theta}$, the phasor $-e^{-i\theta}$ is also shown. That's because the subtraction of $e^{-i\theta}$ from $e^{i\theta}$ is equivalent to the addition of $e^{i\theta}$ and $-e^{-i\theta}$.

The addition of the counter-rotating phasors $e^{i\theta}$ and $-e^{-i\theta}$ at various values of θ is illustrated in Fig. 1.28. At $\theta = 0$, both of these phasors point along the horizontal (real) axis of the complex plane, but the phasor $-e^{-i\theta}$ points along the *negative* real axis while the phasor $e^{i\theta}$ points along the positive real axis. Adding these two opposite-direction phasors produces a zero resultant phasor, which is shown as a dot labeled $e^{i\theta} + (-e^{-i\theta})$. Hence the expression $(e^{i\theta} - e^{-i\theta})/(2i)$ has a magnitude of 0, which is the value of $\sin\theta$ for $\theta = 0$.

Now look at the phasors $e^{i\theta}$ and $-e^{-i\theta}$ for $\theta = 30°$. The phasor $e^{i\theta}$ points above the positive real axis by 30°, and the phasor $-e^{-i\theta}$ points above the negative real axis by 30°. So the positive horizontal (real) component of $e^{i\theta}$ cancels

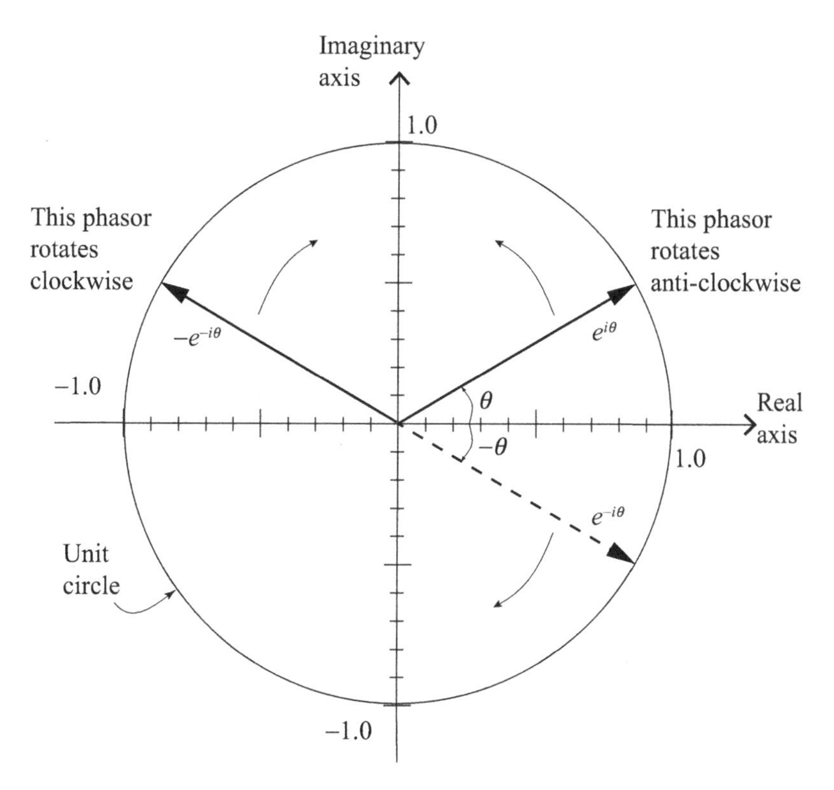

Figure 1.27 Counter-rotating phasors for the sine function.

out the negative horizontal component of $-e^{-i\theta}$, while the positive vertical (imaginary) component of $e^{i\theta}$ adds to the positive vertical component of $-e^{-i\theta}$ to give a resultant that points along the positive imaginary axis with a value of $1.0i$. Dividing this value by $2i$ gives 0.5, which is the value of $\sin\theta$ for $\theta = 30°$.

If you increase θ to 60°, the horizontal (real) components of $e^{i\theta}$ and $-e^{-i\theta}$ still cancel out, while the vertical (imaginary) components add to the larger value of 1.73, and dividing by 2 gives 0.866, the value of $\sin\theta$ for $\theta = 60°$.

For $\theta = 90°$, the vertical (imaginary) components are one for both $e^{i\theta}$ and $-e^{-i\theta}$, and the horizontal (real) components are both zero, so the value of the resultant $e^{i\theta} - e^{-i\theta}$ is $2i$, and dividing by $2i$ gives one, which is the value of $\sin\theta$ for $\theta = 90°$.

The same analysis applies for θ between 90° and 180°; for those angles the resultant phasor $e^{i\theta} - e^{-i\theta}$ returns to zero, as you can see on the right side of Fig. 1.28. As θ increases from 180° to 360°, the $e^{i\theta}$ and $e^{-i\theta}$ phasors continue rotating, and the resultant $e^{i\theta} + e^{-i\theta}$ points along the negative vertical (imaginary) axis, as expected for the sine function at these angles.

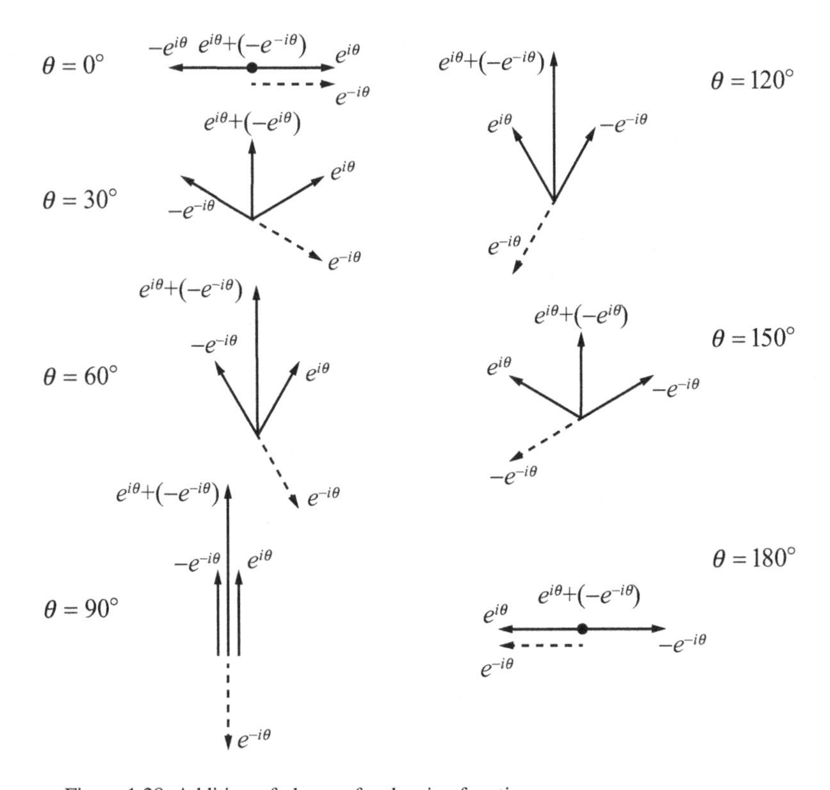

Figure 1.28 Addition of phasors for the sine function.

So the sine function can be represented by the two counter-rotating phasors $e^{i\theta}$ and $-e^{-i\theta}$, and at any angle the addition of those phasors and division by $2i$ yields the value of the sine of that angle.

In some texts, a simplified version of phasor representation of sinusoidal functions is used. In that version, the complex plane is presented simply as a pair of perpendicular axes (typically without the "real" and "imaginary" labels), and the value of the function is taken as the projection of a phasor of length A onto the vertical axis. The angle of the phasor with respect to the positive (rightward) horizontal axis is often labelled ϕ and is given by $\phi = \omega t + \epsilon$. Examples of this use of phasor representation are shown in Fig. 1.29.

Notice that in this figure the phasor is not identified as $e^{i\phi}$, the axes are not identified, and the counter-rotating vectors $e^{i\phi}$ and $e^{-i\phi}$ are not shown. But, at any given angle ϕ, the projection onto the vertical axis gives the same values as the full complex-phasor addition approach.

To see why this works, consider the sine function. In this case, the subtraction of the components of the counter-rotating phasors has two effects:

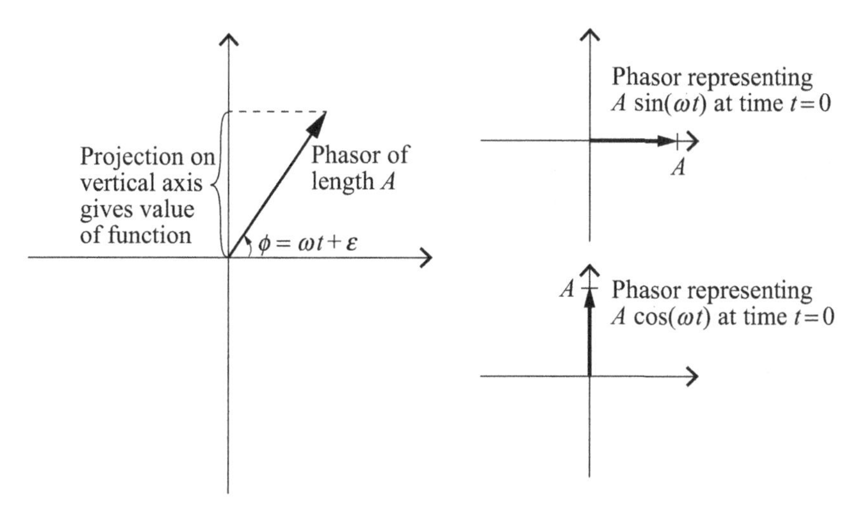

Figure 1.29 A simplified representation of sinusoidal functions.

It cancels out the horizontal (real) component of the resultant; and it doubles the length of the vertical (imaginary) component. So the process of adding the counter-rotating phasors and then dividing by $2i$ gives exactly the same result as finding the projection of a single phasor onto the vertical axis.

In the case of the cosine function, the same analysis applies, as long as you recall that $\cos\phi = \sin(\phi + \pi/2)$ and draw the phasor representing the cosine function along the vertical axis at time $t = 0$.

Using this simplified approach, the two wavefunctions (y_1 and y_2) shown in Figs. 1.23 and 1.24 can be represented by the two phasors shown in Fig. 1.30.

Notice that both of these phasors rotate anti-clockwise at the same rate (since they have the same ω), maintaining the phase difference between them ($\pi/2$ in this case). If you were to add these two phasors together, the resultant would be another phasor rotating at the same frequency and maintaining constant length, but its projection onto the vertical axis would grow larger and smaller (and become negative) as y_1 and y_2 rotate. This summation of two sinusoidal waves is an important concept when you consider the superposition of waves, which you can read about in Section 2.3 in the next chapter.

Another important concept that students sometimes find confusing is the meaning of "negative frequency". The confusion probably arises from the fact that frequency is inversely proportional to period, and period is always a positive number. So how can frequency be negative?

The answer is that a negative value for frequency has meaning only after you've defined anti-clockwise phasor rotation as positive (just as negative velocity has meaning only after you've defined a direction for positive

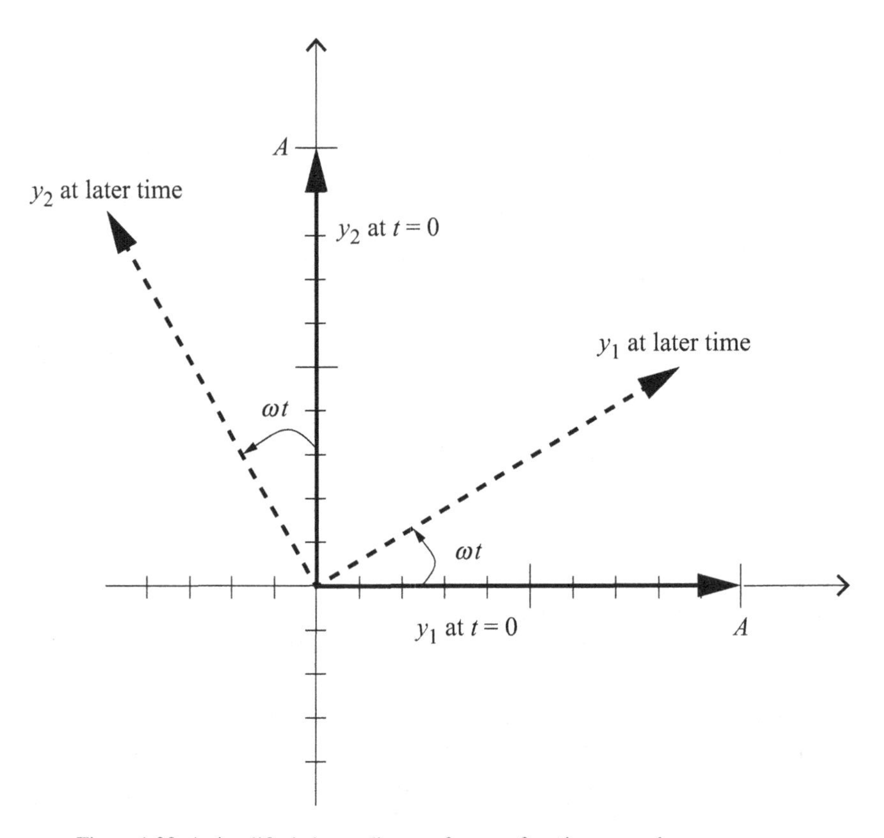

Figure 1.30 A simplified phasor diagram for wavefunctions y_1 and y_2.

velocity). If the angle of anti-clockwise rotation is expressed as ωt, then clockwise rotation must correspond to negative angular frequency ω.

So the counter-rotating phasors that make up the sine and cosine functions can be considered to be one phasor with positive frequency (rotating anti-clockwise) and another phasor with negative frequency (rotating clockwise). This may seem like an esoteric concept, but you'll find it extremely useful if your studies carry you to the land of Fourier analysis.

1.8 Problems

1.1. Find the frequency and angular frequency of the following waves.

(a) A string wave with period of 0.02 s.
(b) An electromagnetic wave with period of 1.5 ns.
(c) A sound wave with period of 1/3 ms.

1.2. Find the period of the following waves.

(a) A mechanical wave with frequency of 500 Hz.
(b) A light wave with frequency of 5.09×10^{14} Hz.
(c) An ocean wave with angular frequency of 0.1 rad/s.

1.3. (a) What is the speed of an electromagnetic wave with wavelength of 2 meters and frequency of 150 MHz?
(b) What is the wavelength of a sound wave with frequency of 9.5 kHz, if the speed of sound is 340 m/s?

1.4. (a) How much does the phase of an electromagnetic wave with frequency of 100 kHz change in 1.5 μs at a fixed location?
(b) What is the difference in phase of a mechanical wave with period of 2 seconds and speed of 15 m/s at two locations separated by 4 meters at some instant?

1.5. If vector $\vec{D} = -5\hat{\imath} - 2\hat{\jmath}$ and vector $\vec{E} = 4\hat{\jmath}$, find the magnitude and direction of the vector $\vec{F} = \vec{D} + \vec{E}$ both graphically and algebraically.

1.6. Verify that each of the complex numbers in Fig. 1.10 has the polar form shown in Fig. 1.12.

1.7. Solve the differential equation $dz/d\theta = iz$ for z.

1.8. Use the power-series representation of $\sin\theta$, $\cos\theta$, and $e^{i\theta}$ to prove the Euler relation $e^{i\theta} = \cos\theta + i\sin\theta$.

1.9. Show that the wavefunction $f(-x-1)$ is shifted in the negative x-direction relative to the wavefunction $f(-x)$.

1.10. Find the phase speed and the direction of propagation of each of the following waves (all units are SI).

(a) $f(x,t) = 5\sin(3x) - t/2$.
(b) $\psi(x,t) = g - 4x - 20t$.
(c) $h(y,t) = 1/[2(2t + x)] + 10$.

2
The wave equation

There are many equations that describe wave behavior and the relationship between wave parameters, and you may well find such equations referred to as "the wave equation". In this chapter, you can read about the most common form of the wave equation, which is a linear, second-order, homogeneous partial differential equation (the meaning of each of these adjectives is explained in Section 2.3 of this chapter). That equation relates the spatial (distance-based) variation of the wavefunction to the temporal (time-based) variation through the wave speed, as described in Section 2.2. Other partial differential equations that are related to the wave equation are discussed in Section 2.4.

If you hope to understand the wave equation (and all other partial differential equations), a good place to begin is to make sure you have a solid understanding of partial derivatives. To help with that, Section 2.1 provides a review of first- and second-order partial derivatives – if you're confident in your understanding of partial derivatives, you can skip that section and jump right into Section 2.2.

2.1 Partial derivatives

If you've taken any course in calculus or college-level physics, you almost certainly encountered ordinary derivatives when you learned how to find the slope of a line ($m = dy/dx$) or how to determine the speed of an object given its position as a function of time ($v_x = dx/dt$). As you probably learned, there are many functions in mathematics and physics that depend on only one independent variable, and ordinary derivatives are all you'll ever need to analyze the changes in such functions.

But, as described in Section 1.6 of Chapter 1, wavefunctions (y) generally depend on two or more independent variables, such as distance (x) and time (t): $y = f(x, t)$. Just as in the case of functions of a single variable, the process of differentiation is very useful in analyzing *changes* in functions of multiple variables. That's exactly where partial derivatives come into play, by extending the concepts of ordinary derivatives to functions of multiple variables. To distinguish between ordinary and partial derivatives, ordinary derivatives are written as d/dx or d/dt and partial derivatives are written as $\partial/\partial x$ or $\partial/\partial t$.

As you may recall, ordinary derivatives allow you to determine the *change* of one variable with respect to another. For example, if you have a function $y = f(x)$, the ordinary derivative of y with respect to x (that is, dy/dx) tells you how much the value of y changes for a small change in the variable x. If you make a graph with y on the vertical axis and x on the horizontal axis, as in Fig. 2.1, then the slope of the line between any two points (x_1, y_1) and (x_2, y_2) on the graph is simply

$$\frac{y_2 - y_1}{x_2 - x_1} = \frac{\Delta y}{\Delta x}.$$

That's because the slope is defined as "the rise over the run", and, since the rise is Δy for a run Δx, the slope of the line between any two points is $\Delta y/\Delta x$.

To precisely represent the slope at a given point on the curve, allow the "run" Δx to become very small. If you write the incremental run as dx and the (also

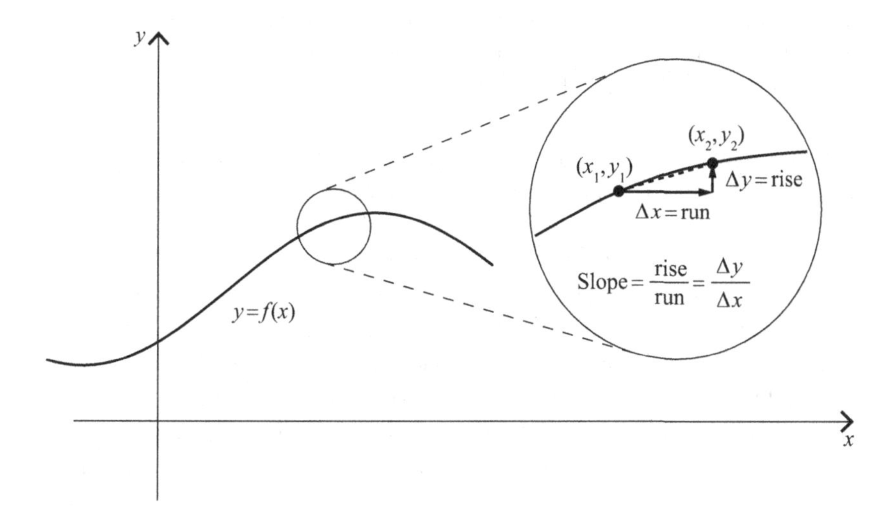

Figure 2.1 The slope of the line $y = f(x)$.

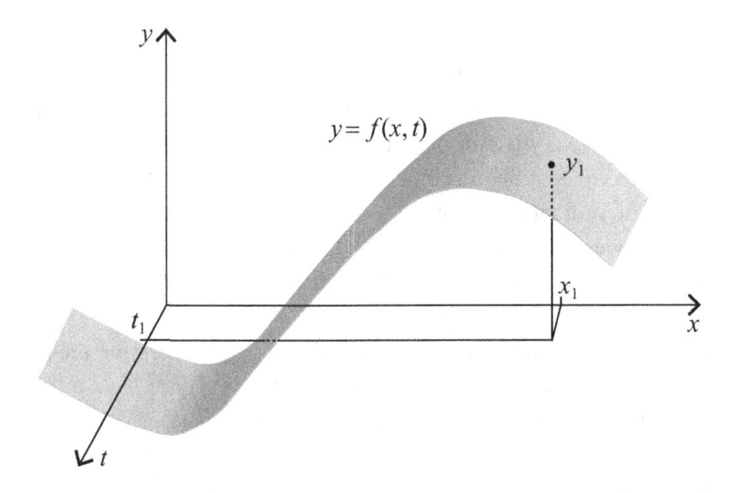

Figure 2.2 The surface in three-dimensional space $y = f(x, t)$.

incremental) rise as dy, then the slope at any point on the line can be written as dy/dx. This is the reasoning that equates the derivative of a function to the slope of the graph of that function.

To extend the process of finding the slope of a function by differentiation to functions such as $y = f(x, t)$, consider a three-dimensional graph of y vs. x and t, such as that shown in Fig. 2.2. The function $y(x, t)$ is shown as a sloping surface in this graph, and the height of the surface above the (x, t) plane is the value of the function y. Since y depends on both x and t, the height of the surface rises and falls as x and t change. And, since the height y may change at a different rate in different directions, a single derivative will not generally suffice to characterize the change in height as you move from one point to another.

Notice that, at the location shown in Fig. 2.3, the slope of the surface is quite steep if you move in the direction of increasing x (while remaining at the same value of t), but the slope is almost zero if you move in the direction of increasing t (while holding your x-value constant).

The occurrence of different slopes in different directions in this figure illustrates the usefulness of *partial* derivatives, which are derivatives formed by allowing one independent variable (such as x or t in Fig. 2.3) to change while holding all other independent variables constant. So the partial derivative $\partial y/\partial x$ represents the slope of the surface at a given location if you move *only along the x-direction* from that location, and the partial derivative $\partial y/\partial t$ represents the slope if you move *only along the t-direction.* These partial derivatives are sometimes written as $\partial y/\partial x|_t$ and $\partial y/\partial t|_x$, in which

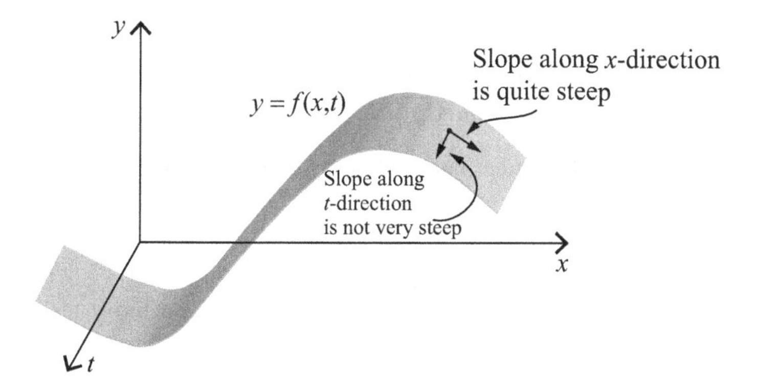

Figure 2.3 Slopes of a surface in three-dimensional space.

the variables that appear in the subscript after the vertical line are held constant.

The good news is that, if you know how to take ordinary derivatives, you already know how to take partial derivatives. Simply treat all variables (with the exception of the one variable over which the derivative is being taken) as constants, and take the derivative as you normally would. You can see how this works in the following example.

Example 2.1 *For the function $y(x, t) = 3x^2 - 5t$, find the partial derivative of y with respect to x and with respect to t.*

To take the partial derivative of y with respect to x, treat t as a constant:

$$\frac{\partial y}{\partial x} = \frac{\partial(3x^2 - 5t)}{\partial x} = \frac{\partial(3x^2)}{\partial x} - \frac{\partial(5t)}{\partial x}$$
$$= 3\frac{\partial(x^2)}{\partial x} - 0 = 6x.$$

For the partial derivative of y with respect to t, treat x as a constant:

$$\frac{\partial y}{\partial t} = \frac{\partial(3x^2 - 5t)}{\partial t} = \frac{\partial(3x^2)}{\partial t} - \frac{\partial(5t)}{\partial t}$$
$$= 0 - 5\frac{\partial t}{\partial t} = -5.$$

Just as you can take "higher-order" ordinary derivatives such as

$$\frac{d}{dx}\left(\frac{dy}{dx}\right) = \frac{d^2y}{dx^2}$$

and

$$\frac{d}{dt}\left(\frac{dy}{dt}\right) = \frac{d^2y}{dt^2},$$

you can also take higher-order partial derivatives. So for example

$$\frac{\partial}{\partial x}\left(\frac{\partial y}{\partial x}\right) = \frac{\partial^2 y}{\partial x^2}$$

tells you the *change* in the x-direction slope of y as you move along the x-direction, and

$$\frac{\partial}{\partial t}\left(\frac{\partial y}{\partial t}\right) = \frac{\partial^2 y}{\partial t^2}$$

tells you the *change* in the t-direction slope as you move along the t-direction.

It's important for you to realize that an expression such as $\partial^2 y/\partial x^2$ is the derivative of a derivative, which is *not the same* as $(\partial y/\partial x)^2$, which is the square of a first derivative (the first is the change in the slope; the second is the square of the slope). By convention the order of the derivative is always written between the "d" or "∂" and the function, as d^2y or $\partial^2 y$, so be sure to look carefully at the location of superscripts when you're dealing with derivatives.

Three-dimensional plots of wavefunctions such as $y(f, t)$ may be considerably more complex than the simple function shown in Figs. 2.2 and 2.3. For example, consider the wavefunction $y(x, t) = A\sin(kx - \omega t)$ plotted in Fig. 2.4.

In this plot, the behavior of the wavefunction y over distance can be seen by looking along the positive x-axis (to the right), and the behavior of y over time can be seen by looking along the positive t-axis (out of the page). Since the x-term and the t-term have opposite signs, this wave is propagating in the positive x-direction; you can see the waveform shifting to the right as time increases (the wave is shown at time $t = 0$ and at three later times).

To help you see the time-behavior of the wave, the black dots labeled 1 through 4 show the value of the wavefunction at position $x = 0$ at four different times (these points all lie in the (y, t) plane, which is shaded in Fig. 2.4). Notice that the shape of the wavefunction from time $t = 0$ to later times (that is, along the t-axis) is a negative sine wave, as expected for the function $y(x, t) = A\sin(kx - \omega t)$ with $x = 0$, since $A\sin(-\omega t) = -A\sin(\omega t)$.

Since the wave equation discussed in the next section involves the partial derivatives of the wavefunction, it may help you to consider the partial

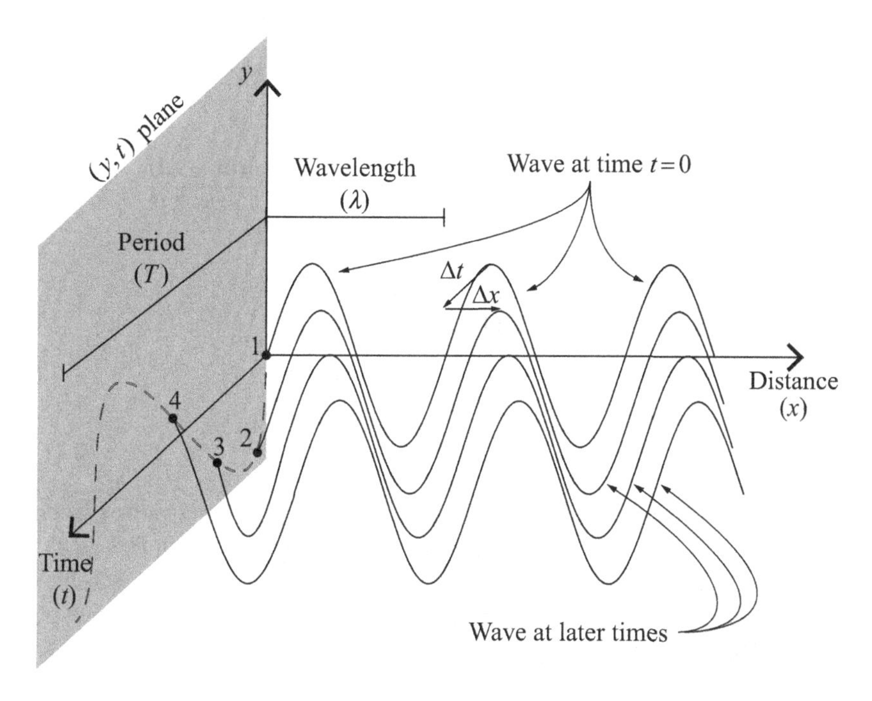

Figure 2.4 A three-dimensional plot of the sinusoidal wave $y(x,t) = A\sin(kx - \omega t)$.

derivatives of $y(x,t) = A\sin(kx - \omega t)$ with constant wavenumber k and angular frequency ω. The first partial derivative of y with respect to x is

$$\begin{aligned}\frac{\partial y}{\partial x} &= \frac{\partial[A\sin(kx-\omega t)]}{\partial x}\\ &= A\frac{\partial[\sin(kx-\omega t)]}{\partial x} = A\cos(kx-\omega t)\frac{\partial(kx-\omega t)}{\partial x}\\ &= A\cos(kx-\omega t)\left[\frac{\partial(kx)}{\partial x} - \frac{\partial(\omega t)}{\partial x}\right] = A\cos(kx-\omega t)\left[k\frac{\partial x}{\partial x} - 0\right]\end{aligned}$$

or

$$\frac{\partial y}{\partial x} = Ak\cos(kx - \omega t). \quad (2.1)$$

The second partial derivative of y with respect to x is

$$\begin{aligned}\frac{\partial^2 y}{\partial x^2} &= \frac{\partial[Ak\cos(kx-\omega t)]}{\partial x}\\ &= Ak\frac{\partial[\cos(kx-\omega t)]}{\partial x} = -Ak\sin(kx-\omega t)\frac{\partial(kx-\omega t)}{\partial x}\\ &= -Ak\sin(kx-\omega t)\left[\frac{\partial(kx)}{\partial x} - \frac{\partial(\omega t)}{\partial x}\right]\\ &= -Ak\sin(kx-\omega t)\left[k\frac{\partial x}{\partial x} - 0\right].\end{aligned}$$

or

$$\frac{\partial^2 y}{\partial x^2} = -Ak^2\sin(kx-\omega t). \tag{2.2}$$

Plots of this wavefunction and its first and second partial derivatives with respect to x at time $t = 0$ are shown in Fig. 2.5. As expected from Eqs. (2.1) and (2.2), the first partial derivative with respect to x has the shape of a cosine function, and the second partial derivative with respect to x has the shape of a negative sine function.

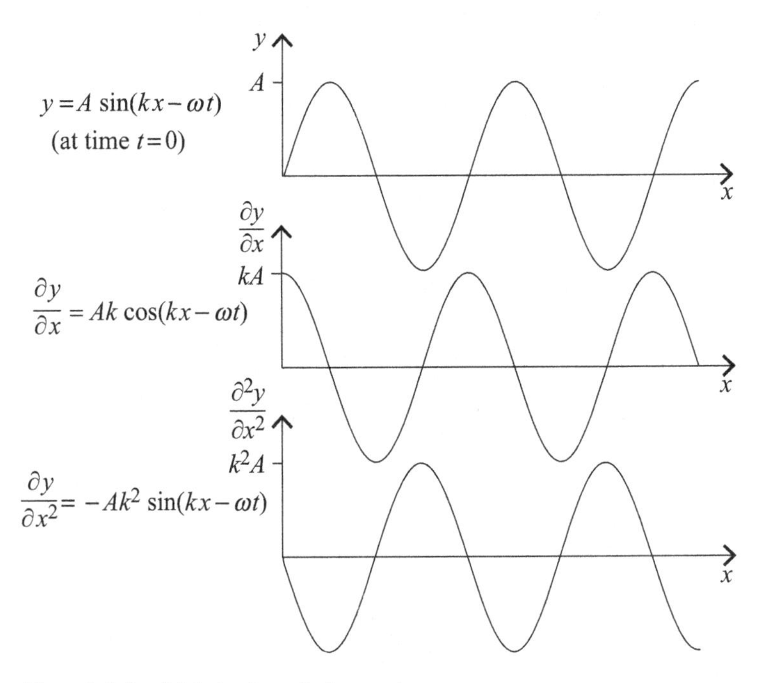

Figure 2.5 Spatial derivatives of a harmonic wave.

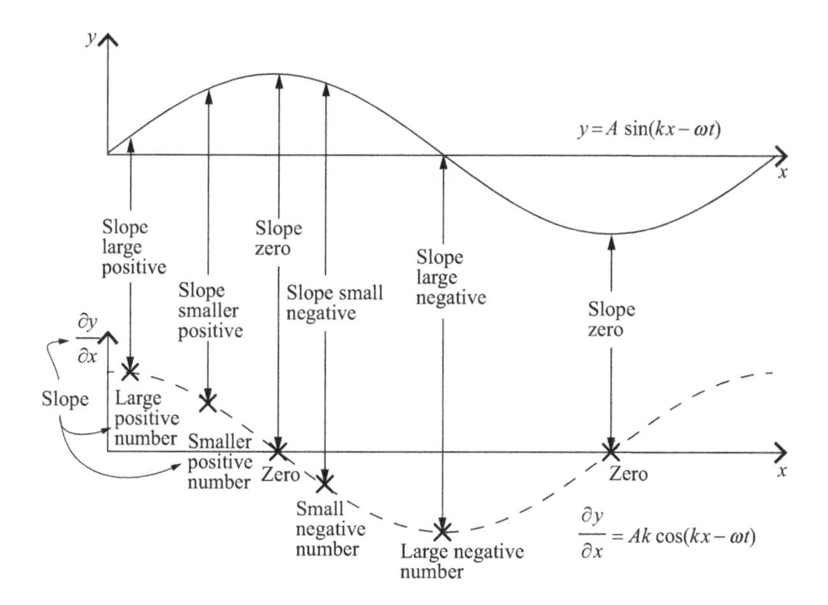

Figure 2.6 The first derivative as slope.

If you're wondering how the cosine shape of the first partial derivatives of $y(x, t)$ relates to the slope of the wavefunction, take a look at Fig. 2.6. The slope of the function y at each value of x becomes the value plotted on the $\partial y/\partial x$ graph for that value of x. And to see how the negative-sine shape of the second partial derivatives of $y(x, t)$ relates to the change of the slope of the wavefunction, take a look at Fig. 2.7. When you estimate the change in the slope of y, remember that the change in slope is negative whenever the slope becomes less positive or when the slope becomes more negative. Likewise, the change in slope is positive whenever the slope becomes more positive or when the slope becomes less negative.

Turning now to the behavior of this wavefunction over time, the first partial derivative of y with respect to t is

$$\begin{aligned}
\frac{\partial y}{\partial t} &= \frac{\partial[A\sin(kx-\omega t)]}{\partial t} \\
&= A\frac{\partial[\sin(kx-\omega t)]}{\partial t} = A\cos(kx-\omega t)\frac{\partial(kx-\omega t)}{\partial t} \\
&= A\cos(kx-\omega t)\left[\frac{\partial(kx)}{\partial t}-\frac{\partial(\omega t)}{\partial t}\right] \\
&= A\cos(kx-\omega t)\left[0-\omega\frac{\partial t}{\partial t}\right]
\end{aligned}$$

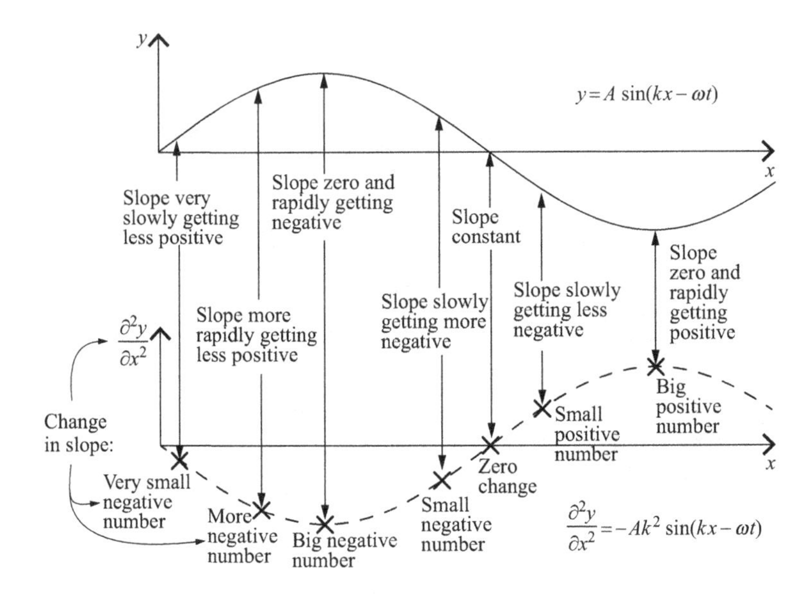

Figure 2.7 The second derivative as the change in slope.

or

$$\frac{\partial y}{\partial t} = -A\omega\cos(kx - \omega t). \tag{2.3}$$

The second partial derivative of y with respect to t is

$$\begin{aligned}\frac{\partial^2 y}{\partial t^2} &= \frac{\partial[-A\omega\cos(kx - \omega t)]}{\partial t}\\ &= -A\omega\frac{\partial[\cos(kx - \omega t)]}{\partial t} = A\omega\sin(kx - \omega t)\frac{\partial(kx - \omega t)}{\partial t}\\ &= A\omega\sin(kx - \omega t)\left[\frac{\partial(kx)}{\partial t} - \frac{\partial(\omega t)}{\partial t}\right]\\ &= A\omega\sin(kx - \omega t)\left[0 - \omega\frac{\partial t}{\partial t}\right].\end{aligned}$$

or

$$\frac{\partial^2 y}{\partial t^2} = -A\omega^2\sin(kx - \omega t). \tag{2.4}$$

So what do these partial derivatives have to do with the subject of this chapter (that is, the wave equation)? As you can see in the next section, the form of the wave equation you're most likely to encounter (the "classical wave equation") is based on the relationship between the second partial derivatives of the wavefunction with respect to time (Eq. (2.4)) and distance (Eq. (2.2)).

2.2 The classical wave equation

In the introduction to this chapter, the most common form of the wave equation is described as a linear, second-order, homogeneous partial differential equation. This is often called the "classical" wave equation, and it usually looks something like this:

$$\frac{\partial^2 y}{\partial x^2} = \frac{1}{v^2}\frac{\partial^2 y}{\partial t^2}. \tag{2.5}$$

There are several different ways to derive this equation. Many authors use the approach of applying Newton's second law to a string under tension, and you can see how that works in Chapter 4. But we think that you can gain a good physical understanding of the wave equation by thinking carefully about the meaning of the partial derivatives that appear on both sides of Eq. (2.5).

Since those second-order partial derivatives involve the change in the slope of the wavefunction $y(x, t)$, it's instructive to consider how the slope of the wavefunction over distance is related to the slope over time under various circumstances. For example, consider the wavefunction $y(x, t) = A\sin(kx - \omega t)$ representing a sinusoidal wave traveling in the positive x-direction. A plot of this wavefunction vs. distance at time $t = 0$ is shown in the top portion of Fig. 2.8, with the slope of $y(x, t)$ between points 1 and 2 shown as the rise over the run $(\Delta y/\Delta x)$.

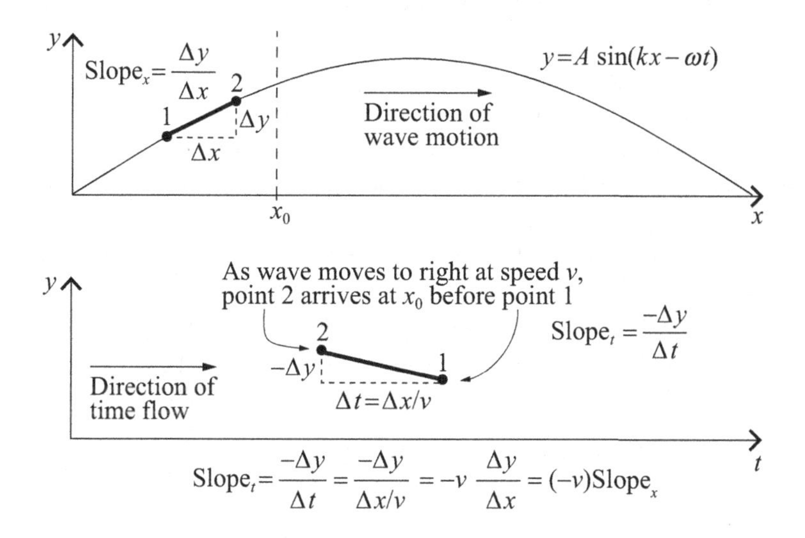

Figure 2.8 Spatial and time slopes of a wave moving in the positive x-direction.

Imagine what you would measure for the value of y (the disturbance produced by the wave) over time if you were sitting at the location x_0. As the wave moves toward positive x (that is, to the right in the distance plot), the portion of the wave called point 2 arrives at your location *before* point 1. So, if you made a plot of the values of y as the wave passes over your location, those values would decrease from the value of point 2 to the value of point 1, as shown in the lower portion of Fig. 2.8. Thus the slope you measure over time (at a given location) has two differences from the slope you measure over distance (at a given time). The first difference is that the magnitude of the slope will be different (since the run in the distance plot is Δx but the run in the time plot is Δt). The second difference is that the sign of slope over distance is positive, but the sign of the slope over time is negative.

Comparing the distance slope to the time slope may seem like an academic exercise, but that comparison leads to a version of the wave equation that is just one step away from the classical wave equation (Eq. (2.5)). To understand how that works, use the fact that the distance increment Δx is related to the time increment Δt by the wave speed v:

$$\Delta x = v\,\Delta t. \tag{2.6}$$

This relationship between Δx and Δt means that the slope over time $(-\Delta y/\Delta t)$ can be written as $-\Delta y/(\Delta x/v) = -v\,\Delta y/\Delta x$. Thus the slope over time is just $-v$ times the slope over distance. Allowing the distance and time increments to shrink toward zero, the Δs become partial derivatives, and the relationship between slopes may be written as

$$\frac{\partial y}{\partial t} = -v\frac{\partial y}{\partial x} \qquad \text{or} \qquad \frac{\partial y}{\partial x} = -\frac{1}{v}\frac{\partial y}{\partial t}. \tag{2.7}$$

This is a perfectly usable first-order wave equation, but it applies only to waves moving in the positive x-direction. To see why that's true, consider the slopes over distance and time for a wave moving in the negative x-direction (to the left in the top plot in Fig. 2.9).

In this case, as the leftward-moving wave passes over an observer at location x_0, the portion of the wave called point 1 arrives before point 2, so the wavefunction value will increase during this interval. This means that a graph of y vs. time will have a positive slope (as does the graph of y vs. distance). Thus the time slope $(\Delta y/\Delta t)$ can be written as $\Delta y/(\Delta x/v) = v\,\Delta y/\Delta x$, and for this wave the slope over time is v times the slope over distance. Again allowing the distance and time increments to shrink toward zero, the relationship between slopes may be written as

$$\frac{\partial y}{\partial t} = v\frac{\partial y}{\partial x} \qquad \text{or} \qquad \frac{\partial y}{\partial x} = \frac{1}{v}\frac{\partial y}{\partial t}. \tag{2.8}$$

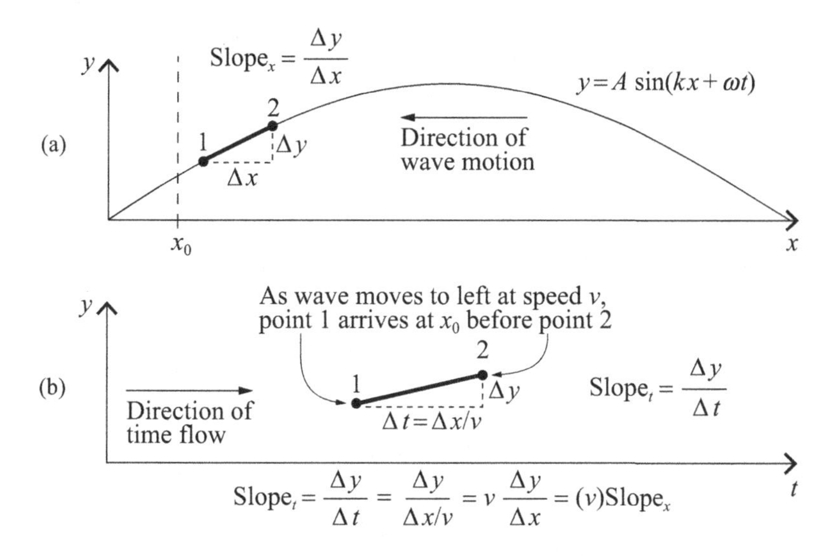

Figure 2.9 Spatial and time slopes of a wave moving in the negative x-direction.

So the first-order wave equation for a wave moving in the negative x-direction (Eq. (2.8)) differs from the first-order wave equation for a wave moving in the positive x-direction (Eq. (2.7)) by a sign. A more useful version of the wave equation would apply to waves moving in either direction, and that's one of the benefits of the second-order classical wave equation. To arrive at that equation, you have to consider not just the slope, but the *change* in the slope of the wavefunction over distance and time.

The change in slope of a wave moving in the positive x-direction is illustrated in Fig. 2.10. As described above, the slope of this wave over distance is positive and the slope over time is negative.

But now take a look at the *change* in the slope over distance and time. Over this interval, as x increases the slope of the wavefunction becomes less positive, so the change in the slope is negative. And as the wave passes over an observer at location x_0, the slope of the wavefunction over time is negative, and it becomes *more negative* during this interval. As the value of the slope becomes more negative, the change in the slope is also negative. So, although the slope over distance and the slope over time have opposite signs for this wave, the *change* in the slope is negative over both time and distance. In this case, relating the distance increment and the time increment using Eq. (2.6) results in two factors of v (one for the slope and one for the change in the slope). Again allowing the distance and time increments to shrink toward zero gives the classical wave equation

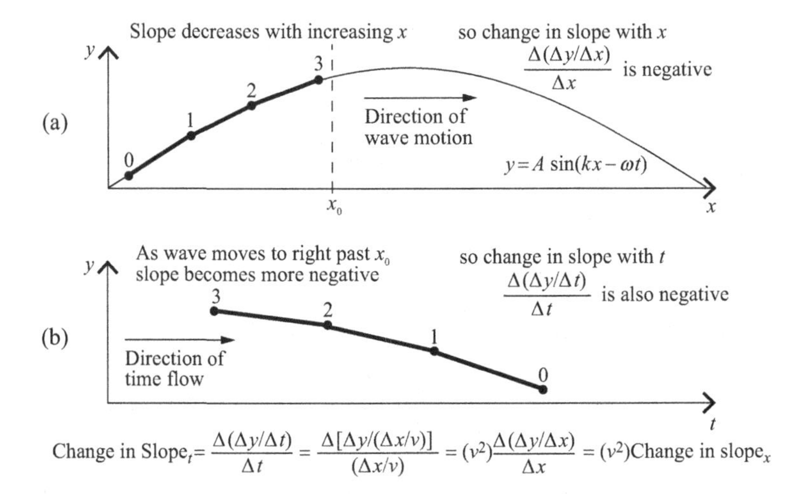

Figure 2.10 Change in slopes for a wave moving in the positive x-direction.

$$\frac{\partial^2 y}{\partial x^2} = \frac{1}{v^2}\frac{\partial^2 y}{\partial t^2}. \tag{2.5}$$

Does this equation also work for waves traveling in the opposite direction (that is, toward negative x)? To see that it does, consider the change in slope over distance and time shown in Fig. 2.11. In this case, the slope over distance and the slope over time are both positive, and the slope becomes less positive over both distance and time, so the change in slope is again negative in both cases. Thus the changes in slope over distance and over time are related by the same factor (positive v^2), and the equation relating the change in slope is again Eq. (2.5).

So while the first-order wave equation applies to waves traveling in one direction, the second-order wave equation takes the same form for waves traveling in the positive x- or in the negative x-direction. Although we used sinusoidal waves to demonstrate this concept, the result is general and applies to waves of any profile.

If this geometric approach to the wave equation isn't your cup of tea, you'll be happy to learn that there's a straightforward route to the first- and second-order wave equation for sinusoidal waves that's based on the derivatives of a wavefunction such as $y(x, t) = A\sin(kx - \omega t)$. From

$$\frac{\partial y}{\partial x} = Ak\cos(kx - \omega t), \tag{2.1}$$

$$\frac{\partial^2 y}{\partial x^2} = -Ak^2\sin(kx - \omega t), \tag{2.2}$$

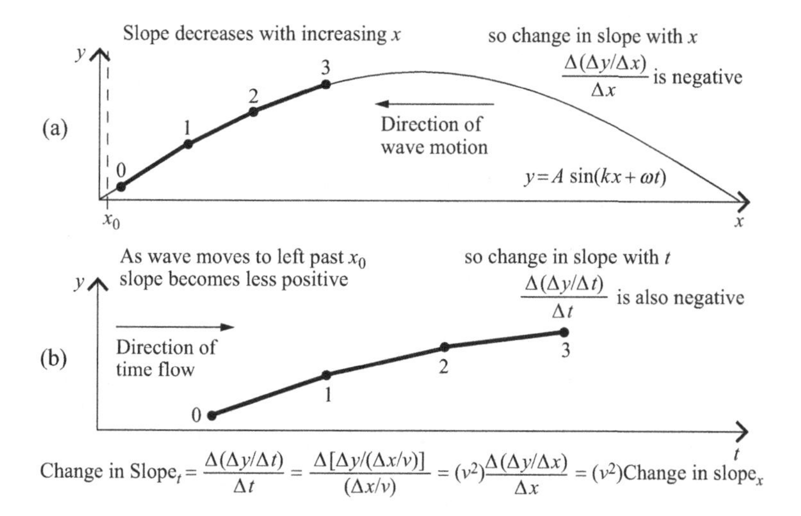

Figure 2.11 Change in slopes for a wave moving in the negative x-direction.

$$\frac{\partial y}{\partial t} = -A\omega \cos(kx - \omega t); \tag{2.3}$$

$$\frac{\partial^2 y}{\partial t^2} = -A\omega^2 \sin(kx - \omega t) \tag{2.4}$$

you can get to the first-order wave equation by solving Eq. (2.3) for $A\cos(kx - \omega t)$

$$A\cos(kx - \omega t) = -\frac{1}{\omega}\frac{\partial y}{\partial t}$$

and then substituting this expression into Eq. (2.1):

$$\frac{\partial y}{\partial x} = Ak\cos(kx - \omega t) = -\frac{k}{\omega}\frac{\partial y}{\partial t}.$$

Using the relation $v = \omega/k$ (Eq. (1.36)) makes this

$$\frac{\partial y}{\partial x} = -\frac{1}{v}\frac{\partial y}{\partial t},$$

which is identical to Eq. (2.7), the first-order wave equation for waves propagating in the positive x-direction. If we had started with the wavefunction $y(x, t) = A\sin(kx + \omega t)$, we would have arrived the first-order wave equation for waves traveling in the negative x-direction (Eq. (2.8)).

Performing a similar analysis on the second-order equations (Eqs. (2.2) and (2.4)) yields the second-order classical wave equation. The first step is to solve Eq. (2.4) for $A\sin(kx - \omega t)$:

$$A\sin(kx-\omega t)=-\frac{1}{\omega^2}\frac{\partial^2 y}{\partial t^2},$$

which you can then substitute into Eq. (2.2):

$$\frac{\partial^2 y}{\partial x^2}=-Ak^2\sin(kx-\omega t)=\frac{k^2}{\omega^2}\frac{\partial^2 y}{\partial t^2}.$$

Again using the relation $v=\omega/k$ (Eq. (1.36)) gives

$$\frac{\partial^2 y}{\partial x^2}=\frac{1}{v^2}\frac{\partial^2 y}{\partial t^2},$$

which is the classical second-order wave equation (Eq. (2.5)). Starting with the wavefunction $y(x,t)=A\sin(kx+\omega t)$ gives the same result.

Once again, although we've used harmonic (sinusoidal) functions to demonstrate this approach, it works just as well for general wavefunctions $f(kx-\omega t)$ and $f(kx+\omega t)$.

Before discussing the properties of the wave equation, we want you to be aware that you may encounter versions of the wave equation that look quite different from the version presented in this section. One common way of writing Eq. (2.2) for harmonic waves is

$$\frac{\partial^2 y}{\partial x^2}=-Ak^2\sin(kx-\omega t)=-k^2y, \tag{2.9}$$

since $y=A\sin(kx-\omega t)$ in this case.

For the same reason, Eq. (2.4) can be written as

$$\frac{\partial^2 y}{\partial t^2}=-A\omega^2\sin(kx-\omega t)=-\omega^2y. \tag{2.10}$$

You may also encounter the "dot" and "double-dot" notation, in which first derivatives with respect to time are signified by a dot over the variable, as in $dx/dt=\dot{x}$ and $\partial y/\partial t=\dot{y}$, and second derivatives with respect to time are signified by two dots over the variable, as in $\partial^2 y/\partial t^2=\ddot{y}$. Using this notation, Eq. (2.10) becomes

$$\ddot{y}=-\omega^2y$$

and the classical wave equation looks like this:

$$\frac{\partial^2 y}{\partial x^2}=\frac{1}{v^2}\ddot{y}.$$

Another common notation for derivatives uses subscripts to indicate the variable with respect to which the derivative is taken. For example, the first partial derivative of y with respect to x may be written as

$$\frac{\partial y}{\partial x} \equiv y_x,$$

where the symbol $\equiv$ means "is defined as". Using this notation, the second partial derivative with respect to t may be written as

$$\frac{\partial^2 y}{\partial t^2} \equiv y_{tt},$$

so the classical wave equation looks like this:

$$y_{xx} = \frac{1}{v^2} y_{tt}.$$

The classical wave equation can be extended to higher dimensions by adding partial derivatives in other directions. For example, for a spatially three-dimensional wavefunction $\Psi(x, y, z, t)$, the classical wave equation is

$$\frac{\partial^2 \Psi}{\partial x^2} + \frac{\partial^2 \Psi}{\partial y^2} + \frac{\partial^2 \Psi}{\partial z^2} = \frac{1}{v^2} \frac{\partial^2 \Psi}{\partial t^2}. \tag{2.11}$$

You may see this written as

$$\nabla^2 \Psi = \frac{1}{v^2} \frac{\partial^2 \Psi}{\partial t^2}, \tag{2.12}$$

in which the symbol ∇^2 represents the Laplacian operator, which is described in Chapter 5.

Whichever form you encounter, remember that the wave equation tells you that the change in the slope of the waveform over distance equals $1/v^2$ times the change in the slope of the wavefunction over time.

2.3 Properties of the wave equation

When you encounter the classical wave equation, it's likely to be accompanied by some or all of the words "linear, homogeneous, second-order partial differential equation". You may also see the word "hyperbolic" included in the list of adjectives. Each of these terms has a very specific mathematical meaning that's an important property of the classical wave equation. But there are versions of the wave equation to which some of these words don't apply, so it's useful to spend some time understanding them.

The "linear" characteristic of the classical wave equation is perhaps the most important, but the discussion of linearity and its implications is somewhat longer than the discussion of the other characteristics. And since the discussion of "partial" is the shortest, we'll start with that and save linearity for last.

Partial. The classical wave equation is a *partial* differential equation (PDE) because it depends on changes in the wavefunction with respect to more than one variable (such as x and t). The alternative is an *ordinary* differential equation (ODE), which depends on changes with respect to only a single variable. An example of the latter is Newton's second law, which states that the acceleration of an object (that is, the second derivative of the object's position with respect to time) is equal to the sum of the external forces on the object (ΣF_{ext}) divided by the object's mass (m). The one-dimensional version is

$$\frac{d^2x}{dt^2} = \frac{\Sigma F_{\text{ext}}}{m}. \tag{2.13}$$

As described in Section 2.1, you can recognize partial differential equations by the presence of the symbol ∂ (as in $\partial/\partial t$) instead of the letter d (as in d/dt). Ordinary differential equations are generally easier to solve than partial differential equations, so one solution technique for certain types of PDEs is to turn them into ODEs. You can see how this works in Section 2.4 of this chapter.

Homogeneous. The classical wave equation is *homogeneous* because it contains only terms that involve the dependent variable (in this case, the displacement y) or derivatives of the dependent variable (such as $\partial y/\partial x$ or $\partial^2 y/\partial t^2$). Mathematically, that means the classical (homogeneous) wave equation looks like

$$\frac{\partial^2 y}{\partial x^2} - \frac{1}{v^2}\frac{\partial^2 y}{\partial t^2} = 0, \tag{2.14}$$

as opposed to the *inhomogeneous* case, which looks like

$$\frac{\partial^2 y}{\partial x^2} - \frac{1}{v^2}\frac{\partial^2 y}{\partial t^2} = F(x,t), \tag{2.15}$$

where $F(x,t)$ represents some function of the independent variables x and t (but not y).

To determine whether a differential equation is homogeneous, just collect all the terms involving the dependent variable $y(x,t)$, including derivatives, on the left side of the equals sign, and gather all terms not involving $y(x,t)$ on the right side. If there are no terms on the right side (that is, if the right side is zero as in Eq. (2.14)), the differential equation is homogeneous. But any term on the right side (that is, any term such as $F(x,t)$ in Eq. (2.15) that does not involve the dependent variable y) makes the equation inhomogeneous (such equations are also called "non-homogeneous").

Don't fall into the trap of just checking whether the differential equation equals zero; after all, you can always move all of the terms (whether they involve the dependent variable or not) to the left side of the equals sign, leaving zero on the right side. When testing for homogeneity, you have to look at the content of each term and move those terms not involving the dependent variable $y(x, t)$ to the right side of the equation.

You may be wondering exactly what the extra function $F(x, t)$ means in inhomogeneous differential equations. Depending on the application, you may see this term called a "source" or an "external force", and those are good clues as to its meaning. The term not involving the dependent variable always represents an external stimulus of some kind. To see that, look back at Eq. (2.13). All terms with the position x and its derivatives are on the left, but there's still the expression $\Sigma F_{\text{ext}}/m$ on the right, so Newton's second law is inhomogeneous.[1] In this case the physical meaning of $F(x, t) = \Sigma F_{\text{ext}}/m$ is clear – it's the total external force on the object divided by the object's mass, which means it must have dimensions of force over mass, with SI units of newtons per kilogram. Nevertheless, you'll see references to this and other functions $F(x, t)$ as "external forces" even in cases in which the functions don't have the dimension of force. Whatever they're called, such terms represent the contribution of an external stimulus.

Second-order. The classical wave equation is a *second-order* partial differential equation because the order of a differential equation is always determined by the highest-order derivative in the equation (and in this case, the time and space derivatives are both second derivatives). Even if an equation is second-order in space but only first-order in time, it's still considered to be a second-order equation. You can see an example of this (the heat equation) in Section 2.4 of this chapter.

Second-order partial differential equations are very common in physics and engineering, perhaps because the change in the change of a quantity is somehow more fundamental than the first-order derivative, or at least more closely related to things we can measure. For example, it's not the change in an object's position (velocity) that's related to the total external force in Newton's second law, it's the change in the object's velocity (acceleration), and acceleration is the second derivative of position. Likewise, in the classical wave equation, it's the change in the waveform's slope over distance that's

[1] There is a homogeneous version of Newton's second law, and it's his first law: $d^2x/dt^2 = 0$. This is the case of zero total external force.

related to the change in the slope over time, and these changes in slope are second derivatives.

Hyperbolic. As mentioned at the start of this section, you may come across a related system of classification of differential equations in which the classical wave equation is called a "hyperbolic" differential equation. As you may have learned in a geometry class, hyperbolas are a form of conic section (along with ellipses and parabolas) that can be represented by simple equations. And it turns out that the classical wave equation has a similar form to the equation for a hyperbola:

$$\frac{y^2}{a^2} - \frac{x^2}{b^2} = 1, \tag{2.16}$$

in which the constants a and b determine the "flatness" of the hyperbola.

To compare this with the classical wave equation (Eq. (2.5)), it helps to first get both terms in the classical wave equation onto the left side:

$$\frac{\partial^2 y}{\partial x^2} - \frac{1}{v^2}\frac{\partial^2 y}{\partial t^2} = 0. \tag{2.17}$$

This helps, but Eq. (2.16) still doesn't look very much like the classical wave equation. The trick is to think of these two equations as analogous: The differential second-order term $\partial^2 y/\partial x^2$ appears in the same location as the algebraic second-order term $(y/a)^2$, and the differential second-order term $(1/v^2)(\partial^2 y/\partial t^2)$ appears in the same location as the algebraic second-order term $(x/b)^2$. It's important that you remember that the second partial derivative is *not* the same as the square of the first partial derivative; "second-order" means "taking a second derivative" in differential equations, and "second-order" means "the square or product of two first-order variables" in algebraic equations. With that caveat, you can say that both the classical wave equation and the equation for a hyperbola involve the difference between two second-order terms. Hence it's the second derivatives in the wave equation, as well as the negative sign between them, that make the wave equation "hyperbolic".

At this point, you may be concerned that the right side of the equation for a hyperbola is one while the right side of the classical wave equation is zero. But if you consider the equation

$$\frac{y^2}{a^2} - \frac{x^2}{b^2} = 0 \tag{2.18}$$

you find that the solutions are two straight lines that cross at the origin, which is a special case of a hyperbola. So even the homogeneous version of the wave equation does fit the description of "hyperbolic". And, as you can see in Section 2.4, there are useful differential equations with a first-order time derivative and a second-order space derivative, and such equations are characterized as "parabolic".

Linear. The classical wave equation is *linear* because all of the terms involving the wavefunction $y(x, t)$ and derivatives of $y(x, t)$ are raised to the first power, and there are no cross terms involving the product of the wavefunction and its derivatives. Linear differential equations may contain second-order (and higher-order) derivatives, because (as explained in Section 2.1) $\partial^2 y/\partial x^2$ represents the change in the slope of y with x, which is not the same as $(\partial y/\partial x)^2$. If a differential equation does include terms with higher powers or cross terms of the wavefunction and its derivatives, that differential equation is said to be *nonlinear*.

An extremely powerful characteristic of all linear differential equations, including the classical wave equation, is that solutions obey the "superposition principle". This principle describes what happens when two or more waves occupy the same space at the same time. This is a situation in which the behavior of waves is very different from the behavior of solid objects. When solid objects attempt to occupy the same space at the same time, they experience a collision, which generally alters their motion or shape, but each object tends to retain its identity. In contrast, when two linear waves simultaneously occupy the same place, their displacements from equilibrium combine to produce a new wave that also satisfies the wave equation. Neither wave is destroyed in this process, although only the combined wave is observable during the interaction. If the waves then propagate away from the region of overlap, the original characteristics of each wave are again observable. So, unlike particles, waves may "pass through" one another rather than colliding, producing a new wave while overlapping (you can find a more thorough comparison of particles and waves in Chapter 6).

The superposition principle explains why this happens. Mathematically, the superposition principle says that, if two wavefunctions $y_1(x, t)$ and $y_2(x, t)$ are each a solution to the linear wave equation, then their sum at every point in space and time, $y_{\text{total}}(x, t) = y_1(x, t) + y_2(x, t)$, is also a solution. You can prove this for two waves traveling at the same speed v by writing the wave equation of each wave:

$$\begin{aligned}\frac{\partial^2 y_1(x,t)}{\partial x^2} - \frac{1}{v^2}\frac{\partial^2 y_1(x,t)}{\partial t^2} &= 0,\\ \frac{\partial^2 y_2(x,t)}{\partial x^2} - \frac{1}{v^2}\frac{\partial^2 y_2(x,t)}{\partial t^2} &= 0,\end{aligned} \tag{2.19}$$

and then adding these two equations:

$$\frac{\partial^2 y_1(x,t)}{\partial x^2} + \frac{\partial^2 y_2(x,t)}{\partial x^2} - \frac{1}{v^2}\frac{\partial^2 y_1(x,t)}{\partial t^2} - \frac{1}{v^2}\frac{\partial^2 y_2(x,t)}{\partial t^2} = 0.$$

This can be simplified to

$$\frac{\partial^2 [y_1(x,t) + y_2(x,t)]}{\partial x^2} - \frac{1}{v^2}\frac{\partial^2 [y_1(x,t) + y_2(x,t)]}{\partial t^2} = 0$$

and, since $y_{\text{total}}(x,t) = y_1(x,t) + y_2(x,t)$, this is

$$\frac{\partial^2 y_{\text{total}}(x,t)}{\partial x^2} - \frac{1}{v^2}\frac{\partial^2 y_{\text{total}}(x,t)}{\partial t^2} = 0. \tag{2.20}$$

Thus the result of adding two (or more) waves that satisfy the wave equation is another wave that also satisfies the wave equation. And, if you're wondering whether this works when the waves are traveling at different speeds, the answer is that it does, and you can read about the effects of different speeds in Section 3.4 of Chapter 3.

The following example shows the superposition principle in action.

Example 2.2 *Consider two sine waves with the following wavefunctions:*

$$y_1(x,t) = A_1 \sin(k_1 x + \omega_1 t + \epsilon_1),$$
$$y_2(x,t) = A_2 \sin(k_2 x + \omega_2 t + \epsilon_2).$$

If these two waves have the same amplitude, $A_1 = A_2 = A = 1$, the same wavenumber, $k_1 = k_2 = k = 1$ rad/m, and the same angular frequency, $\omega_1 = \omega_2 = \omega = 2$ rad/s, but the first wave $y_1(x,t)$ has a phase constant $\epsilon_1 = 0$ and the second wave $y_2(x,t)$ has a phase constant of $\epsilon_2 = +\pi/3$, determine the charateristics of the wave that results from the addition of these waves.

Since the distance term and the time term have the same sign for both of these waves, you know that both waves are traveling in the negative x-direction, and since the wave phase speed $v = \omega/k$ (see Eq. (1.36)), they also have the same speed. By comparing the phase constants for the two waves, you also know that $y_2(x,t)$ leads $y_1(x,t)$ by a phase difference of $\pi/3$ (if you don't recall why more-positive phase constant results in a leading wave in this case, look back to Section 1.6 of Chapter 1). Inserting the values given above, the two wavefunctions may be written as

$$\begin{aligned} y_1(x,t) &= A_1 \sin(k_1 x + \omega_1 t + \epsilon_1) = \sin(x + 2t + 0), \\ y_2(x,t) &= A_2 \sin(k_2 x + \omega_2 t + \epsilon_2) = \sin(x + 2t + \pi/3) \end{aligned} \tag{2.21}$$

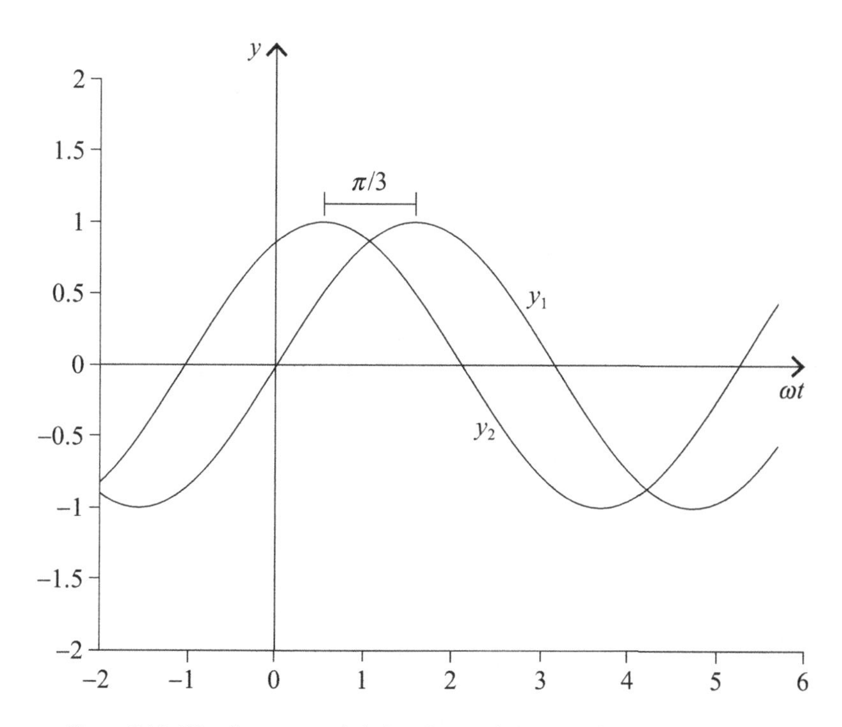

Figure 2.12 Waveforms $y_1 = \sin(\omega t)$ and $y_2 = \sin(\omega t + \pi/3)$ at location $x = 0$.

and, at $x = 0$, they look like Fig. 2.12.

To understand how these two waves add to produce a new wave, take a look at Fig. 2.13. In this figure, the graphical addition of the two waves is shown to result in another sinusoidal wave, drawn with a dashed line. This resultant wave has the same frequency as the two original waves, but it has a different phase constant and larger amplitude.

Some algebra can show the same result, starting with the expression for the resultant wave:

$$y_{\text{total}}(x, t) = \sin(x + 2t) + \sin(x + 2t + \pi/3). \tag{2.22}$$

A useful trigonometric identity here is

$$\sin\theta_1 + \sin\theta_2 = 2\sin\left(\frac{\theta_1 + \theta_2}{2}\right)\cos\left(\frac{\theta_1 - \theta_2}{2}\right), \tag{2.23}$$

and plugging in $\theta_1 = x + 2t$ and $\theta_2 = x + 2t + \pi/3$ gives

$$y_{\text{total}}(x, t) = 2\sin\left(\frac{2(x + 2t) + \pi/3}{2}\right)\cos\left(\frac{-\pi/3}{2}\right). \tag{2.24}$$

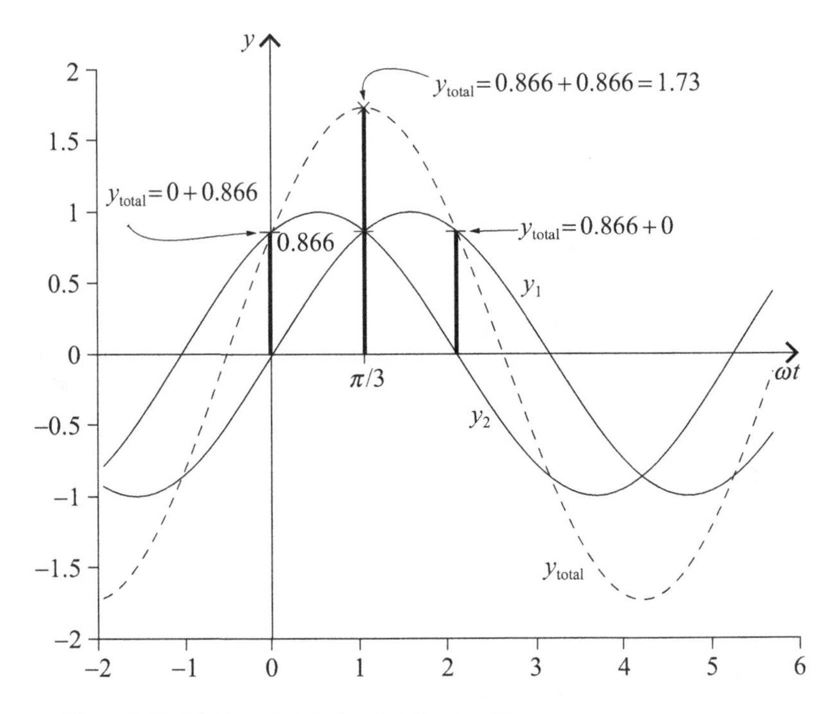

Figure 2.13 Addition of $\sin(\omega t)$ and $\sin(\omega t + \pi/3)$.

The sine term in this expression simplifies to $\sin(x + 2t + \pi/6)$, which is a wave with wavenumber $k = 1$ rad/m and angular frequency $\omega = 2$ rad/s (hence the same wavelength and frequency as the original waves), but with a phase constant $\epsilon = \pi/6$ (in this case, the average of the original phases of zero and $\pi/3$). What about the amplitude? The rest of Eq. (2.24) gives $A = 2\cos(-\pi/6) \approx 1.73$. So the amplitude is larger than the original $A = 1$, but not twice as large (since in this case the two original waves don't reach their peak values at the same time).

Another very powerful way to analyze the superposition of waves is through the use of phasors. Using the simplified phasor approach described in Section 1.7, each of the waves used in the previous example can be represented by a rotating phasor, as shown in Fig. 2.14 for the location $x = 0$ at time $t = 0$. Remember that in this simplified approach the value of the wavefunction at any time is given by the projection of the rotating phasor onto the vertical axis, so at the instant shown the values of $y_1(x, t)$ and $y_2(x, t)$ are given by

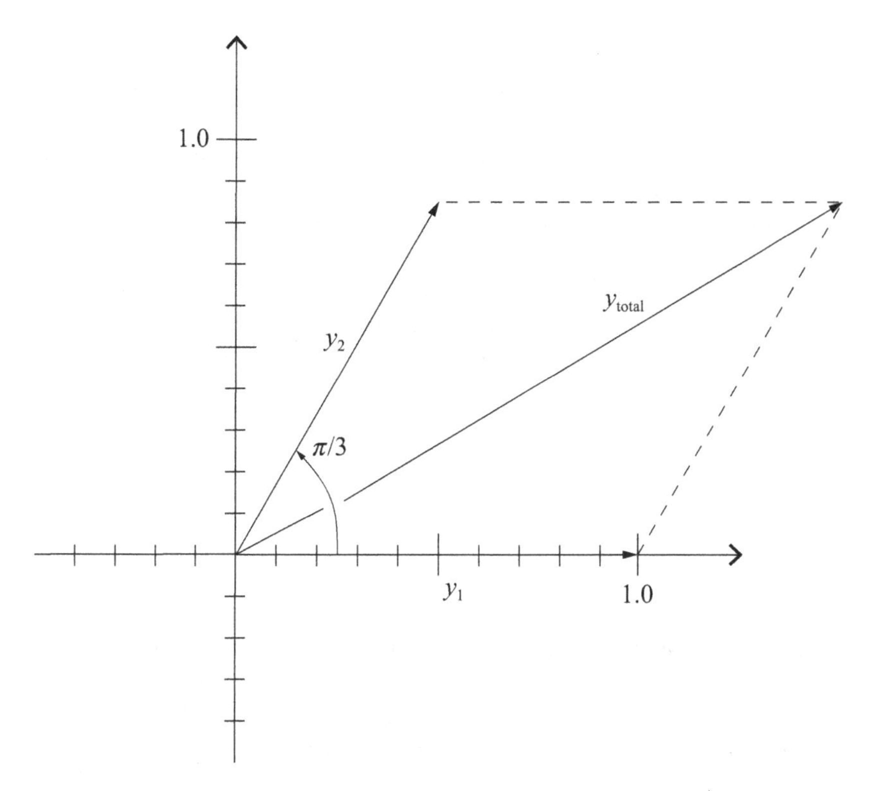

Figure 2.14 Wave addition by phasors.

$$y_1(0,0) = A_1 \sin[k_1 x + \omega_1 t + \epsilon_1] = (1)\sin[(1)(0) + (2)(0) + 0] = 0,$$
$$y_2(0,0) = A_2 \sin[k_2 x + \omega_2 t + \epsilon_2] = (1)\sin[(1)(0) + (2)(0) + \pi/3]$$
$$= 0.866.$$

The power of the phasor approach becomes evident when you consider how easy it is to determine the resultant of adding these two phasors. To find the amplitude and direction of the phasor representing the combination of waves $y_1(x, t)$ and $y_2(x, t)$, simply perform a vector addition of the two phasors representing the waves. Several ways to perform vector addition are described in Section 1.3; the graphical "parallelogram" approach is shown in Fig. 2.14. A quick look at the resultant y_{total} shows that it has magnitude almost twice as large as the magnitude of y_1 or y_2 and that the phase constant of the resultant is halfway between the zero phase constant of y_1 and the $\pi/3$ phase constant of y_2. A ruler and protractor reveal that the length of y_{total} is in fact 1.73, the phase constant is $\pi/6$, and the projection of the y_{total} phasor onto the vertical axis gives a value of 0.866, as found by the algebraic wave-addition approach discussed previously.

If you prefer to use the "addition of components" approach to finding the resultant phasor y_{total}, you can use the geometry of Fig. 2.14 to see that y_1 has an x-component of 1 and a y-component of zero, while y_2 has an x-component of $1\cos(\pi/3) = 0.5$ and a y-component of $1\sin(\pi/3) = 0.866$. Adding the x-components of y_1 and y_2 gives the x-component of y_{total} as 1.5, and adding the y-components of y_1 and y_2 gives the y-component of y_{total} as 0.866. Thus the magnitude and phase angle of y_{total} are

$$A_{\text{total}} = \sqrt{(1.5^2 + 0.866^2)} = 1.73, \tag{2.25}$$

$$\epsilon_{\text{total}} = \tan^{-1}\left(\frac{0.866}{1.5}\right) = \pi/6 \tag{2.26}$$

and the length of the projection of the y_{total} phasor onto the vertical axis is $1.73\sin(\pi/6) = 0.866$, as expected for $y_{\text{total}}(0, 0)$.

So the use of phasors makes wave addition straightforward, but you may be concerned that we have performed this addition only at the specific time we selected ($t = 0$). But since both phasors are rotating at the same rate ωt (since the waves have the same angular frequency), the angle between the y_1 phasor and the y_2 phasor will remain the same. That means that the magnitude of the

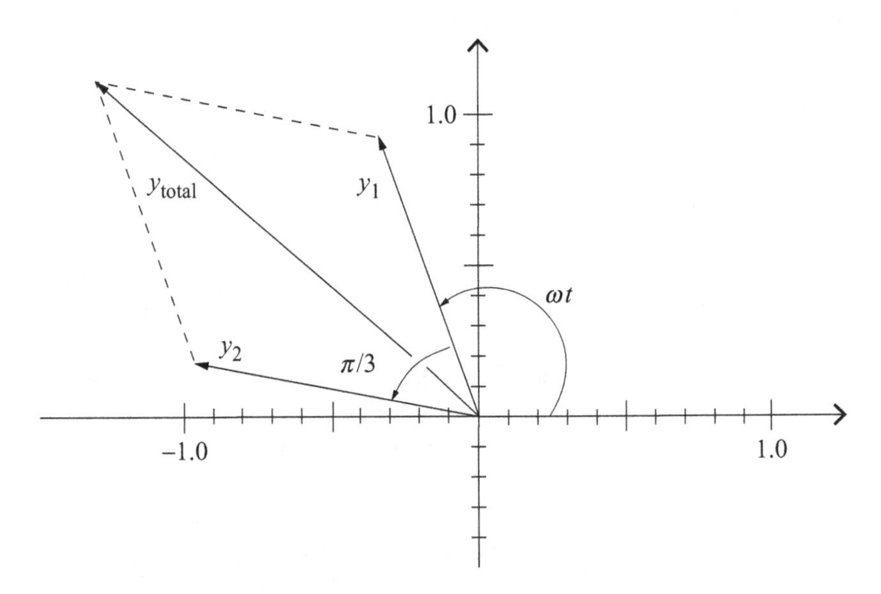

Figure 2.15 Wave addition by phasors at a later time.

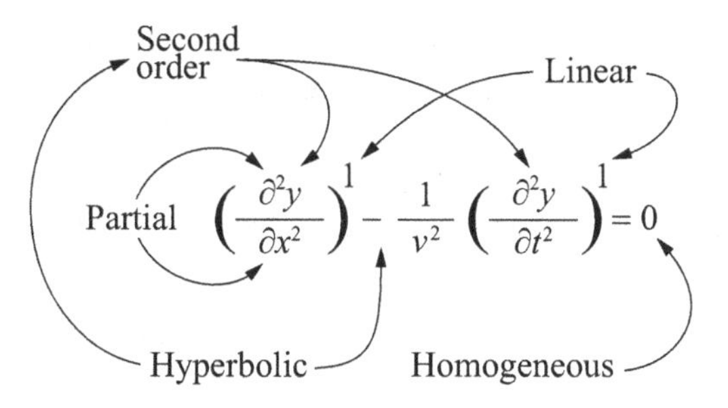

Figure 2.16 The expanded wave equation.

resultant phasor y_{total} will not change over time, although its projection onto the vertical axis will change, as indicated in Fig. 2.15.

Summary. All of the characteristics of the classical wave equation discussed in this section are summarized in the expanded equation shown in Fig. 2.16.

2.4 PDEs related to the wave equation

The classical wave equation is second order, linear, and hyperbolic, but there are other partial differential equations that pertain to motion through space and time, and those equations have some characteristics that are similar and some that are different from the characteristics of the classical wave equation. In this section you'll find a short discussion of several of these equations, since they involve concepts and techniques that are helpful in understanding solutions to the wave equation.

The advection equation. Unlike the second-order classical wave equation, the advection equation is a first-order wave equation. In fact, you've already seen it in Section 2.2: It's the one-way wave equation:

$$\frac{\partial y(x,t)}{\partial x} = -\frac{1}{v}\frac{\partial y(x,t)}{\partial t}. \tag{2.27}$$

What use is a one-way wave equation? "Advection" is a kind of transport mechanism (that is, a way for material or energy to be transferred from one place to another) specifically describing the way substances move when they are carried along in a current. If you have, for example, pollutants in a river or pollen in an air current, the advection equation can model the movement more simply than the classical wave equation.

The Korteweg–de Vries equation. Not all wave equations are linear; a well-known example of a nonlinear wave equation is the Korteweg–de Vries (KdV) equation, which describes, for example, small-amplitude, shallow, and confined water waves, called "solitary waves" or "solitons". A soliton looks like a wave pulse that retains its shape as it travels. The KdV equation has the form

$$\frac{\partial y(x,t)}{\partial t} - 6y(x,t)\frac{\partial y(x,t)}{\partial x} + \frac{\partial^3 y(x,t)}{\partial x^3} = 0. \tag{2.28}$$

The nonlinear term is the middle one, $6y(\partial y/\partial x)$, because it includes the product of two different terms involving $y(x,t)$. Because it's a nonlinear equation, the solutions no longer obey the superposition principle: During an interaction, the total amplitude is not the sum of the original amplitudes. However, once the solitons have passed through each other, they return to their original shapes.

The heat equation. While the classical wave equation is hyperbolic, the heat equation is parabolic, having a form analogous to $y = ax^2$. In other words, while it's still second-order in space (and hence a second-order PDE), it's only first-order in time:

$$\frac{\partial y(x,t)}{\partial t} = a\frac{\partial^2 y(x,t)}{\partial x^2}, \tag{2.29}$$

where a is the thermal diffusivity, a measure of how easily heat transfers through a system. You won't usually see the heat equation classified as a wave equation, although it describes the behavior of disturbances in time and space. That's due to the behavior of the solution, which is dissipative rather than oscillatory. But wave packets can also dissipate in time, so there's overlap between these cases.

How can you tell that the solutions to the heat equation dissipate in time? By examining the time-dependent portion of the solution, which you can do through a common method called "separation of variables". The assumption behind this method is that, although the solution $y(x,t)$ depends on both space and time, the behavior in time $T(t)$ is independent of the behavior in space $X(x)$. That is, the wavefunction $y(x,t)$ can be written as the product of one term $T(t)$ that depends only on time and another term $X(x)$ that depends only on location:

$$y(x,t) = T(t)X(x). \tag{2.30}$$

Why this form and not, say, $y(x,t) = T(t) + X(x)$? The answer is that Eq. (2.30) works for many physically meaningful situations. So the problem of determining the evolution of the wavefunction over time reduces to finding an expression for the time-dependent term $T(t)$.

To see how that works, plug Eq. (2.30) into the heat equation:

$$\frac{\partial[T(t)X(x)]}{\partial t} = a\frac{\partial^2[T(t)X(x)]}{\partial x^2}. \tag{2.31}$$

Because $X(x)$ is a constant with respect to time, it's unaffected by $\partial/\partial t$ and comes out of the time derivative; similarly, $T(t)$ is independent of space and so is unaffected by $\partial/\partial x$. Pulling these functions out of the derivatives leaves

$$X(x)\frac{\partial T(t)}{\partial t} = aT(t)\frac{\partial^2 X(x)}{\partial x^2}. \tag{2.32}$$

The next step is to isolate all functions and derivatives of t on one side and x on the other. In this case it's as simple as dividing both sides by $X(x)$ and $T(t)$. In fact, when reading about the separation of variables method you'll often see authors explicitly divide the equation by $y(x,t) = T(t)X(x)$ at this step.

The heat equation now looks like

$$\frac{1}{T(t)}\frac{\partial T(t)}{\partial t} = a\frac{1}{X(x)}\frac{\partial^2 X(x)}{\partial x^2}. \tag{2.33}$$

This equation may not look particularly useful, but take a step back and consider each side. The left side depends only on time (t) and does not vary with location (x). The right side depends only on location and does not vary with time. But if this equation is true at every location at every time, then neither side can vary at all (if it did, then, as time passes at a fixed location, the left side would vary but the right side would not). Thus both sides must be constant, and since the sides equal one another, they must equal the same constant. If you call that constant $-b$, then Eq. (2.33) becomes

$$\frac{1}{T(t)}\frac{dT(t)}{dt} = -b, \tag{2.34}$$

which is an ordinary differential equation. It's also separable, meaning that T and t can be put on different sides of the equation. Writing $T(t)$ as T for simplicity and multiplying both sides by dt gives

$$\frac{1}{T}dT = -b\,dt.$$

Integrating both sides gives

$$\int \frac{dT}{T} = \int -b\,dt$$

or

$$\ln T = -bt + c,$$

in which c is the combined integration constant for both sides. In order to solve for $T(t)$, you'll need to apply the inverse function of ln, the natural log, which is the exponential: $e^{\ln T} = T$. Doing so gives

$$T(t) = e^{-bt+c} = e^{-bt}e^{c} = Ae^{-bt}. \tag{2.35}$$

In this expression, the constant term e^{c} has been absorbed into A. It's the last term (e^{-bt}) that makes the solution dissipative: Whatever the behavior in space, the total solution will decay exponentially as e^{-bt}, with the rate of decay determined by the constant b.

The Schrödinger equation. The Schrödinger equation bears a stronger resemblance to the heat equation than to the classical wave equation, but as you can see in Chapter 6, its solutions definitely have the characteristics of waves. Like the heat equation, the Schrödinger equation has a first-order derivative with respect to time and a second-order derivative with respect to position. However, it has an additional factor of i with the time derivative, and that factor has a significant impact on the nature of the solutions. You can see that by considering this form of the Schrödinger equation:

$$i\hbar\frac{\partial y(x,t)}{\partial t} = -\frac{\hbar^2}{2m}\frac{\partial^2 y(x,t)}{\partial t^2} + Vy(x,t). \tag{2.36}$$

In this equation, V is the potential energy of the system and $\hbar$ is the reduced Planck constant (see Chapter 6).

Just as with the heat equation, the time behavior of the solutions can be found by using the separation of variables. Assuming that the overall solution has the form of Eq. (2.30), the Schrödinger equation becomes

$$i\hbar\frac{\partial T(t)X(x)}{\partial t} = -\frac{\hbar^2}{2m}\frac{\partial^2 T(t)X(x)}{\partial t^2} + VT(t)X(x). \tag{2.37}$$

Pulling $T(t)$ out of the space derivatives and $X(x)$ out of the time derivatives and dividing by $T(t)X(x)$ gives

$$\frac{i\hbar}{T(t)}\frac{\partial T(t)}{\partial t} = -\frac{\hbar^2}{2mX(x)}\frac{\partial^2 X(x)}{\partial t^2} + V. \tag{2.38}$$

Using the same reasoning as described above gives a time-only equation of

$$\frac{i\hbar}{T(t)}\frac{dT(t)}{dt} = E. \tag{2.39}$$

This constant is the energy of the state, which is why it's called E. This is now an ordinary differential equation and can be arranged into

$$\frac{dT}{T} = -\frac{iE}{\hbar}\,dt.$$

Now integrate each side,

$$\ln T = -\frac{iEt}{\hbar} + c,$$

and solve for $T(t)$:

$$T(t) = Ae^{-iEt/\hbar}. \tag{2.40}$$

As explained in Section 1.5 of Chapter 1, unlike the decaying exponential function e^{-x}, the real and imaginary parts of e^{ix} are oscillatory. Thus the solutions to the Schrödinger equation are very different from the solutions to the heat equation.

2.5 Problems

2.1. Find $\partial f/\partial x$ and $\partial f/\partial t$ for the function $f(x,t) = 3x^2t^2 + \frac{1}{2}x + 3t^3 + 5$.

2.2. For the function $f(x,t)$ of Problem 2.1, find $\partial^2 f/\partial x^2$ and $\partial^2 f/\partial t^2$.

2.3. For the function $f(x,t)$ of Problem 2.1, show that $\partial^2 f/\partial x\,\partial t$ gives the same result as $\partial^2 f/\partial t\,\partial x$.

2.4. Does the function $Ae^{i(kxt-\omega t)}$ satisfy the classical wave equation? If so, prove it. If not, say why not.

2.5. Does the function $A_1 e^{i(kx+\omega t)} + A_2 e^{i(kx-\omega t)}$ satisfy the classical wave equation? If so, prove it. If not, say why not.

2.6. Does the function $Ae^{(ax+bt)^2}$ satisfy the classical wave equation? If so, what is the speed of the wave described by this function?

2.7. Sketch the solutions to the hyperbolic equation $y^2/a^2 - x^2/b^2 = 1$ for various values of a and b.

2.8. Make time-domain plots of $y_1(x,t) = A\sin(kx + \omega t + \epsilon_1)$ and $y_2(x,t) = A\sin(kx + \omega t + \epsilon_2)$ and their sum for $A = 1$ m, $k = 1$ rad/m, and $\omega = 2$ rad/s at positions $x = 0.5$ m and $x = 1.0$ m over at least one full period of oscillation. Take $\epsilon_1 = 1.5$ rad and $\epsilon_2 = 0$.

2.9. Sketch the phasors for the waveforms of the previous problem (and their sum) at $x = 1$ m at times $t = 0.5$ s and $t = 1.0$ s.

2.10. Does the function $Ae^{i(kx-\omega t)}$ satisfy the advection equation as given in Eq. (2.27)? What about the function $Ae^{i(kx+\omega t)}$?

3
Wave components

Before diving into mechanical, electromagnetic, and quantum waves, it will help you to understand the general solutions to the wave equation (Section 3.1) and the importance of boundary conditions to those solutions (Section 3.2). And, although single-frequency waves provide a useful introduction to many of the concepts important to wave theory, the waves you're likely to encounter in real-world applications often include multiple frequency components. Putting those components together to produce the resultant wave is the subject of Fourier synthesis, and determining the amplitude and phase of the individual components can be accomplished through Fourier analysis. With an understanding of the basics of Fourier theory (Section 3.3), you'll be ready to deal with the important topics of wave packets and dispersion (Section 3.4). As in other chapters, the discussion of these topics is modular, so you should feel free to skip any of the sections of this chapter with which you're already familiar.

3.1 General solutions to the wave equation

In seeking a general solution to the classical one-dimensional wave equation

$$\frac{\partial^2 y}{\partial x^2} = \frac{1}{v^2}\frac{\partial^2 y}{\partial t^2} \tag{3.1}$$

you're likely to encounter an approach developed by the French mathematician Jean le Rond d'Alembert in the eighteenth century. To understand d'Alembert's solution, begin by defining two new variables involving both x and t

$$\xi = x - vt,$$
$$\eta = x + vt$$

and consider how to write the wave equation using these variables. Since the wave equation involves second derivatives in both space and time, start with the chain rule:

$$\frac{\partial y}{\partial x} = \frac{\partial y}{\partial \xi}\frac{\partial \xi}{\partial x} + \frac{\partial y}{\partial \eta}\frac{\partial \eta}{\partial x}.$$

But $\partial \xi/\partial x$ and $\partial \eta/\partial x$ both equal one, so

$$\frac{\partial y}{\partial x} = \frac{\partial y}{\partial \xi}(1) + \frac{\partial y}{\partial \eta}(1) = \frac{\partial y}{\partial \xi} + \frac{\partial y}{\partial \eta}.$$

Taking a second derivative with respect to x gives

$$\begin{aligned}
\frac{\partial^2 y}{\partial x^2} &= \frac{\partial}{\partial x}\left(\frac{\partial y}{\partial \xi} + \frac{\partial y}{\partial \eta}\right) \\
&= \frac{\partial}{\partial \xi}\left(\frac{\partial y}{\partial \xi} + \frac{\partial y}{\partial \eta}\right)\frac{\partial \xi}{\partial x} + \frac{\partial}{\partial \eta}\left(\frac{\partial y}{\partial \xi} + \frac{\partial y}{\partial \eta}\right)\frac{\partial \eta}{\partial x} \\
&= \left(\frac{\partial^2 y}{\partial \xi^2} + \frac{\partial^2 y}{\partial \xi\,\partial \eta}\right)(1) + \left(\frac{\partial^2 y}{\partial \eta\,\partial \xi} + \frac{\partial^2 y}{\partial \eta^2}\right)(1).
\end{aligned}$$

But, as long as the function y has continuous second derivatives, the order of differentiation in mixed partials is irrelevant, so

$$\frac{\partial^2 y}{\partial \xi\,\partial \eta} = \frac{\partial^2 y}{\partial \eta\,\partial \xi},$$

and

$$\frac{\partial^2 y}{\partial x^2} = \frac{\partial^2 y}{\partial \xi^2} + 2\frac{\partial^2 y}{\partial \xi\,\partial \eta} + \frac{\partial^2 y}{\partial \eta^2}. \tag{3.2}$$

Applying the same procedure to the time derivative of y gives

$$\frac{\partial y}{\partial t} = \frac{\partial y}{\partial \xi}\frac{\partial \xi}{\partial t} + \frac{\partial y}{\partial \eta}\frac{\partial \eta}{\partial t}. \tag{3.3}$$

In this case $\partial \xi/\partial t = -v$ and $\partial \eta/\partial t = +v$, so

$$\frac{\partial y}{\partial t} = \frac{\partial y}{\partial \xi}(-v) + \frac{\partial y}{\partial \eta}(v) = -v\frac{\partial y}{\partial \xi} + v\frac{\partial y}{\partial \eta}.$$

Taking a second derivative with respect to t gives

$$\begin{aligned}
\frac{\partial^2 y}{\partial t^2} &= \frac{\partial}{\partial t}\left(-v\frac{\partial y}{\partial \xi} + v\frac{\partial y}{\partial \eta}\right) \\
&= \frac{\partial}{\partial \xi}\left(-v\frac{\partial y}{\partial \xi} + v\frac{\partial y}{\partial \eta}\right)\frac{\partial \xi}{\partial t} + \frac{\partial}{\partial \eta}\left(-v\frac{\partial y}{\partial \xi} + v\frac{\partial y}{\partial \eta}\right)\frac{\partial \eta}{\partial t} \\
&= \left(-v\frac{\partial^2 y}{\partial \xi^2} + v\frac{\partial^2 y}{\partial \xi\,\partial \eta}\right)(-v) + \left(-v\frac{\partial^2 y}{\partial \eta\,\partial \xi} + v\frac{\partial^2 y}{\partial \eta^2}\right)(v).
\end{aligned}$$

Thus

$$\frac{\partial^2 y}{\partial t^2} = v^2 \frac{\partial^2 y}{\partial \xi^2} - 2v^2 \frac{\partial^2 y}{\partial \xi \, \partial \eta} + v^2 \frac{\partial^2 y}{\partial \eta^2}. \tag{3.4}$$

With the second partial derivatives of y with respect to x and t in hand, you're ready to insert these expressions (Eqs. (3.2) and (3.4)) into the wave equation (Eq. (3.1)):

$$\frac{\partial^2 y}{\partial \xi^2} + 2\frac{\partial^2 y}{\partial \xi \, \partial \eta} + \frac{\partial^2 y}{\partial \eta^2} = \frac{1}{v^2}\left(v^2 \frac{\partial^2 y}{\partial \xi^2} - 2v^2 \frac{\partial^2 y}{\partial \xi \, \partial \eta} + v^2 \frac{\partial^2 y}{\partial \eta^2}\right)$$

or

$$\left(\frac{\partial^2 y}{\partial \xi^2} - \frac{\partial^2 y}{\partial \xi^2}\right) + \left(2\frac{\partial^2 y}{\partial \xi \, \partial \eta} + 2\frac{\partial^2 y}{\partial \xi \, \partial \eta}\right) + \left(\frac{\partial^2 y}{\partial \eta^2} - \frac{\partial^2 y}{\partial \eta^2}\right) = 0.$$

This shows that the change of variables from x and t to ξ and η has simplified the classical wave equation to this:

$$\frac{\partial^2 y}{\partial \xi \, \partial \eta} = 0. \tag{3.5}$$

Although this equation may not immediately give you much physical insight, it can be integrated to reveal a great deal about the nature of the solutions to the wave equation. To see that, begin by writing this equation as

$$\frac{\partial}{\partial \xi}\left(\frac{\partial y}{\partial \eta}\right) = 0$$

and think about what this means. If the partial derivative with respect to ξ of $\partial y/\partial \eta$ is zero, then $\partial y/\partial \eta$ cannot depend on ξ. That means that $\partial y/\partial \eta$ must be a function of η alone, so you can write

$$\frac{\partial y}{\partial \eta} = F(\eta), \tag{3.6}$$

where F represents the function of η that describes how y changes with η. This equation can be integrated to give

$$y = \int F(\eta) d\eta + \text{constant}, \tag{3.7}$$

where "constant" means any function that doesn't depend on η (so it could be any function of ξ). Calling this function $g(\xi)$, Eq. (3.7) becomes

$$y = \int F(\eta) d\eta + g(\xi), \tag{3.8}$$

and, letting the integral of $F(\eta)$ equal another function of η called $f(\eta)$, you have

$$y = f(\eta) + g(\xi) \tag{3.9}$$

or

$$y = f(x + vt) + g(x - vt). \tag{3.10}$$

This is the general solution to the classical one-dimensional wave equation, and it tells you that every wavefunction $y(x, t)$ that satisfies the wave equation can be interpreted as the sum of two waves propagating in opposite directions with the same speed. That's because $f(x + vt)$ represents a disturbance moving in the negative x-direction at speed v and $g(x - vt)$ represents a disturbance moving in the positive x-direction also at speed v.

Of course, like most general solutions, Eq. (3.10) gives you the big picture of what's happening, but you're going to need additional information if you want to find the values of $y(x, t)$ at specific times and places. That information usually comes if the form of boundary conditions, which are the subject of the next section. But if you'd like to see a quick illustration of how d'Alembert's solution to the wave equation can be used, take a look at the following example.

Example 3.1 *If the functions f and g in Eq. (3.10) both represent sine waves of amplitude A, how does the wavefunction $y(x, t)$ behave?*

To answer this question, write f and g as

$$f(x + vt) = A\sin(kx + \omega t),$$
$$g(x - vt) = A\sin(kx - \omega t)$$

(if you're concerned that v doesn't appear explicitly on the right side of these equations, recall from Chapter 1 that $kx - \omega t$ can be written as $k(x - \omega/kt)$, and $\omega/k = v$, where v is the phase velocity of the wave).

Inserting these expressions for f and g into the general solution for the wave equation (Eq. (3.10)) gives

$$y = f(x + vt) + g(x - vt)$$
$$= A\sin(kx + \omega t) + A\sin(kx - \omega t).$$

But

$$\sin(kx + \omega t) = \sin(kx)\cos(\omega t) + \cos(kx)\sin(\omega t)$$

and

$$\sin(kx - \omega t) = \sin(kx)\cos(\omega t) - \cos(kx)\sin(\omega t),$$

so

$$\begin{aligned} y &= A[\sin(kx)\cos(\omega t) + \cos(kx)\sin(\omega t)] \\ &\quad + A[\sin(kx)\cos(\omega t) - \cos(kx)\sin(\omega t)] \\ &= A[\sin(kx)\cos(\omega t) + \sin(kx)\cos(\omega t) + \cos(kx)\sin(\omega t) \\ &\qquad - \cos(kx)\sin(\omega t)] \\ &= 2A\sin(kx)\cos(\omega t). \end{aligned}$$

This expression for the wavefunction y may look superficially like a traveling wave (it does, after all, involve both kx and ωt), but notice that the kx and the ωt appear in *different* terms, and this has a profound effect on the behavior of the wave over space and time.

You can see that behavior by plotting y over a range of x values at several different times. As you can see in Fig. 3.1, the wavefunction $y(x, t)$ is sinusoidal over space and oscillates in time, but the peaks, troughs, and nulls do not move along the x-axis as they would for a propagating wave. Instead,

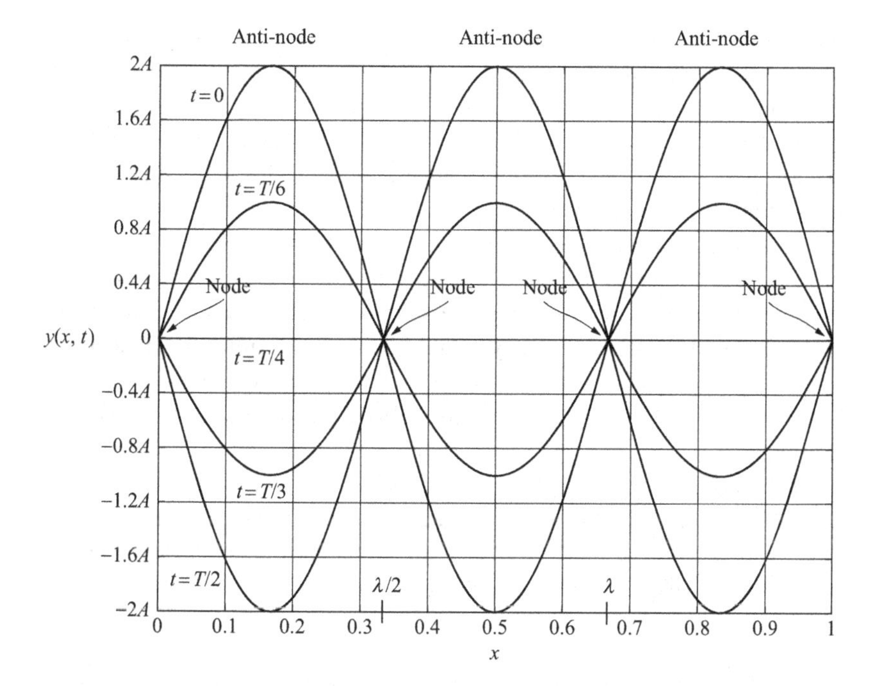

Figure 3.1 $y(x, t)$ over a distance of 1.5 wavelengths at five different times.

the locations of the peaks and nulls remain the same (given by the peaks and nulls of the $\sin(kx)$-term), but the size of the peaks varies over time (due to the $\cos(\omega t)$-term). This cosine term repeats itself over time period T, where $T = 2\pi/\omega$, and the sine term repeats itself over a spatial interval λ, where $\lambda = 2\pi/k$.

So, although this wavefunction is the sum of two propagating waves, the resultant is a non-propagating wave called a "standing wave". As indicated in Fig. 3.1, the locations of the nulls are called the "nodes" of the standing wave, and the locations of the peaks and troughs are called the "anti-nodes" of the standing wave.

How might two such equal-amplitude, counter-propagating waves come into existence? One possibility is that a single wave propagating in one direction reflects from a fixed barrier, and the reflected wave combines with the original wave to produce a standing wave. The fixed barrier in this case is an example of a boundary condition, which you can read about in the next section.

3.2 Boundary conditions

Why are boundary conditions important in wave theory? One reason is this: Differential equations, by their very nature, tell you about the *change* in a function (or, if the equation involves second derivatives, about the change in the change of the function). Knowing how a function changes is very useful, and may be all you need in certain problems. But in many problems you wish to know not only how the function changes, but also what value the function takes on at certain locations or times.

That's where boundary conditions come in. A boundary condition "ties down" a function or its derivative to a specified value at a specified location in space or time.[1] By constraining the solution of a differential equation to satisfy the boundary condition(s), you may be able to determine the value of the function or its derivatives at other locations. We say "may" because boundary conditions that are not well-posed may provide insufficient or contradictory information.

A subset of boundary conditions consists of the "initial conditions", which specify the value of the function or its derivative at the start time of the period under consideration (usually taken as $t = 0$) or at a lower spatial boundary of the region over which the differential equation applies.

[1] "Dirichlet" boundary conditions specify the value of the function itself,"Neumann" boundary conditions specify the value of the function's derivative, and "Cauchy" (or "mixed") boundary conditions combine both Dirichlet and Neumann boundary conditions.

To see how boundary conditions work, consider d'Alembert's general solution to the wave equation (Eq. (3.10)). In some problems, you're given the initial displacement as $I(x) = y(x, 0)$ and the initial (transverse[2]) velocity as $V(x) = \partial y(x, t)/\partial t|_{t=0}$. You can use these conditions and d'Alembert's general solution to determine the value of the function $y(x, t)$ at any location (x) and time (t).

To do that, begin by setting $t = 0$ in the general solution (Eq. (3.10)):

$$y(x, t)|_{t=0} = f(x + vt)|_{t=0} + g(x - vt)|_{t=0}. \tag{3.11}$$

This can be written as

$$y(x, 0) = f(x) + g(x) = I(x), \tag{3.12}$$

where $I(x)$ is the initial displacement at each value of x. Again using the variables $\eta = x + vt$ and $\xi = x - vt$ and then taking the derivative with respect to time and setting $t = 0$ gives

$$\left.\frac{\partial y(x,t)}{\partial t}\right|_{t=0} = \left.\frac{\partial f}{\partial \eta}\frac{\partial \eta}{\partial t}\right|_{t=0} + \left.\frac{\partial g}{\partial \xi}\frac{\partial \xi}{\partial t}\right|_{t=0}.$$

It's important to note that time is set to zero *after* the derivative is taken in this equation. That means you can write $\partial\eta/\partial t = v$ and $\partial\xi/\partial t = -v$, as well as $\partial f/\partial\eta = \partial f/\partial x$ and $\partial g/\partial\xi = \partial g/\partial x$. Thus the initial condition for transverse velocity is

$$\left.\frac{\partial y(x,t)}{\partial t}\right|_{t=0} = \frac{\partial f}{\partial x}v - \frac{\partial g}{\partial x}v = V(x),$$

in which $V(x)$ is a function that specifies the initial velocity at each location x. Thus

$$\frac{\partial f}{\partial x} - \frac{\partial g}{\partial x} = \frac{1}{v}V(x).$$

Integrating this equation over x allows you to write the integral of $V(x)$ in terms of $f(x)$ and $g(x)$ as

$$f(x) - g(x) = \frac{1}{v}\int_0^x V(x)dx, \tag{3.13}$$

where $x = 0$ represents an arbitrary starting location that will turn out not to affect the solution. You already know from Eq. (3.12) that the initial displacement (I) can be written in terms of $f(x)$ and $g(x)$:

$$f(x) + g(x) = I(x), \tag{3.12}$$

[2] The tranverse velocity is not the phase velocity of the wave; in transverse mechanical waves it is the velocity of the particles of the medium moving perpendicular to the direction in which the wave is propagating (see Chapter 4).

so you can isolate $f(x)$ by adding Eq. (3.13) to Eq. (3.12):

$$2f(x) = I(x) + \frac{1}{v}\int_0^x V(x)dx$$

or

$$f(x) = \frac{1}{2}I(x) + \frac{1}{2v}\int_0^x V(x)dx. \tag{3.14}$$

Likewise, you can isolate $g(x)$ by subtracting Eq. (3.13) from Eq. (3.12):

$$2g(x) = I(x) - \frac{1}{v}\int_0^x V(x)dx$$

or

$$g(x) = \frac{1}{2}I(x) - \frac{1}{2v}\int_0^x V(x)dx. \tag{3.15}$$

What good has all this manipulation done? Well, now you have both $f(x)$ and $g(x)$ in terms of the initial conditions for displacement $I(x)$ and transverse velocity $V(x)$. And, if you replace x by $\eta = x + vt$ in the equation for $f(x)$, you get

$$f(\eta) = f(x + vt) = \frac{1}{2}I(x + vt) + \frac{1}{2v}\int_0^{x+vt} V(x + vt)dx.$$

Now replace x by $\xi = x - vt$ in the equation for $g(x)$:

$$g(\xi) = g(x - vt) = \frac{1}{2}I(x - vt) - \frac{1}{2v}\int_0^{x-vt} V(x - vt)dx.$$

With these two equations, you can write the solution $y(x, t)$ over all space and time in terms of the initial conditions for displacement (I) and transverse velocity (V):

$$\begin{aligned} y(x, t) &= f(x + vt) + g(x - vt) \\ &= \frac{1}{2}I(x + vt) + \frac{1}{2v}\int_0^{x+vt} V(x + vt)dx \\ &\quad + \frac{1}{2}I(x - vt) - \frac{1}{2v}\int_0^{x-vt} V(x - vt)dx, \end{aligned}$$

which you can simplify by using the minus sign in front of the last term to switch the limits of integration and combining the integrals:

$$y(x, t) = \frac{1}{2}I(x + vt) + \frac{1}{2}I(x - vt) + \frac{1}{2v}\int_{x-vt}^{x+vt} V(z)dz, \tag{3.16}$$

in which z is a dummy variable (that is, a variable that gets integrated out and can therefore have any name you choose).

Equation (3.16) is d'Alembert's general solution to the wave equation with initial conditions $I(x)$ and $V(x)$. So, if you know $I(x)$ and $V(x)$, you can find the solution $y(x, t)$ at any location (x) and time (t) using this equation.

Whenever you encounter an equation such as Eq. (3.16), you should attempt to understand its physical significance. In this case, the first two terms are simply half the initial displacement (that is, the wave profile) moving toward $-x$ and half moving toward $+x$.

The meaning of the third term is somewhat less apparent, but, by considering the limits of integration, you should be able to see that this is the *accumulated disturbance* at any location (x). To understand why that's true, consider the significance of the interval from $x - vt$ to $x + vt$: This is the distance range over which the disturbance, traveling at speed v, has had time to reach position x. So the integral of the initial velocity function V over that range of x tells you how much of the disturbance has "built up" at any x location.

The following example shows how you can use Eq. (3.16) to find $y(x, t)$ given $I(x)$ and $V(x)$.

Example 3.2 *Find y(x,t) for a wave with the initial displacement condition*

$$y(x, 0) = I(x) = \begin{cases} 5\,[1 + x/(L/2)] & \text{for } -L/2 < x < 0, \\ 5\,[1 - x/(L/2)] & \text{for } 0 < x < L/2, \\ 0 & \text{elsewhere} \end{cases}$$

and initial transverse velocity condition

$$\left.\frac{\partial y(x, t)}{\partial t}\right|_{t=0} = 0.$$

Since you're given the initial displacement (I) and transverse velocity (V) functions, you can use Eq. (3.16) to find $y(x, t)$. But it's often helpful to begin by plotting the initial displacement function, as in Fig. 3.2. In this case, the initial transverse velocity is zero, so there's no need to plot that function.

Now that you have an idea of what the initial displacement looks like, you're ready to use Eq. (3.16):

$$\begin{aligned} y(x, t) &= \frac{1}{2}I(x - vt) + \frac{1}{2}I(x + vt) + \frac{1}{2v}\int_{x-vt}^{x+vt} V(z)dz \\ &= \frac{1}{2}\,[I(x - vt) + I(x + vt)] + 0. \end{aligned}$$

This is just the initial shape of the wave ($I(x)$) scaled by 1/2 and propagating both in the negative and in the positive x-direction while maintaining its shape

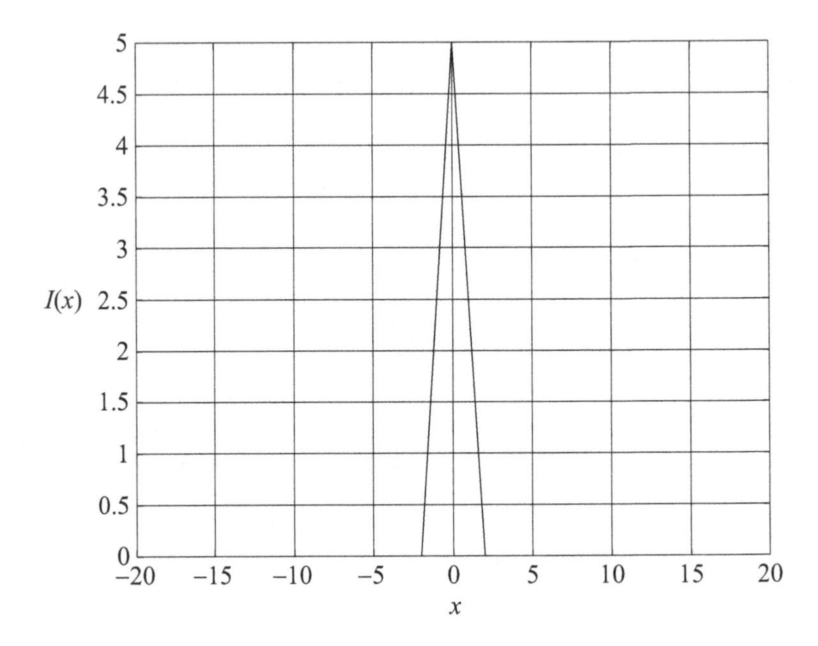

Figure 3.2 The initial displacement $I(x)$ with $L = 4$.

over time, as you can see in Fig. 3.3. In this figure, the tall triangle centered on $x = 0$ is the sum of $\frac{1}{2}[I(x - vt)]$ and $\frac{1}{2}[I(x + vt)]$ at time $t = 0$, which is just $I(x)$. At a later time $t = t_1$, the wavefunction $I(x - vt)$ has propagated a distance vt_1 to the right (toward positive x), while the counter-propagating wavefunction $I(x + vt)$ has moved the same distance to the left (toward negative x), so the two component wavefunctions no longer overlap. As time progresses, the two component wavefunctions continue to move apart, as can be seen by examining the plots for $t = t_2$.

This example illustrated the use of d'Alembert's solution to the wave equation to determine $y(x, t)$ if you're given the initial conditions (displacement $I(x)$ and transverse velocity $V(x)$ at time $t = 0$). As mentioned at the start of this section, the initial conditions are a subset of the boundary conditions, which specify the wavefunction or its derivative at the "boundaries" of the problem. Those "boundaries" may be temporal (time equals zero or infinity, for example), but they can also be spatial ($x = 0$ and $x = L$, for example).

To see how to use such general boundary conditions to find solutions to the wave equation, you first need to understand the separation of variables technique introduced in the discussion of the heat equation in Section 2.4. As described in that section, when you employ this technique, you begin by

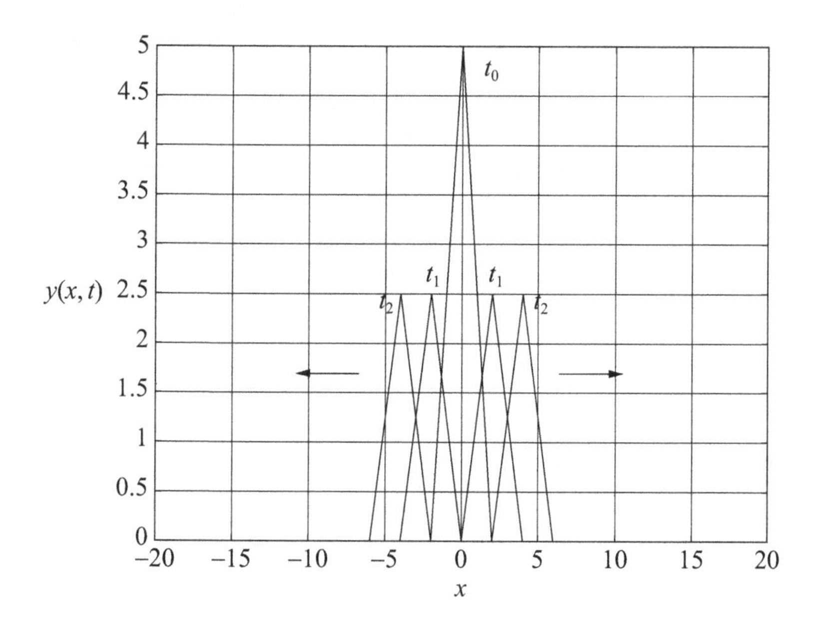

Figure 3.3 Counter-propagating components of a triangle wave at time $t = 0$ and two later times.

assuming that the solution to a partial differential equation involving two or more variables (such as x and t in the one-dimensional wave equation) can be written as the product of functions involving only one variable. So in the case of the one-dimensional wave equation, the solution $y(x, t)$ is assumed to be the product of a function $X(x)$ that depends only on x and another function $T(t)$ that depends only on t. Thus $y(x, t) = X(x)T(t)$, and the classical wave equation

$$\frac{\partial^2 y}{\partial x^2} = \frac{1}{v^2}\frac{\partial^2 y}{\partial t^2}$$

becomes

$$\frac{\partial^2[X(x)T(t)]}{\partial x^2} = \frac{1}{v^2}\frac{\partial^2[X(x)T(t)]}{\partial t^2}. \tag{3.17}$$

But the time function $T(t)$ has no x-dependence and the spatial function $X(x)$ has no time dependence, so T comes out of the first derivative and X comes out of the second:

$$T(t)\frac{\partial^2 X(x)}{\partial x^2} = \frac{1}{v^2}X(x)\frac{\partial^2 T(t)}{\partial t^2}. \tag{3.18}$$

The next step is to divide both sides by the product of the terms $(X(x)T(t))$:

$$\frac{1}{X(x)}\frac{\partial^2 X(x)}{\partial x^2} = \frac{1}{v^2}\frac{1}{T(t)}\frac{\partial^2 T(t)}{\partial t^2}. \tag{3.19}$$

Notice that the left side depends only on x and the right side depends only on t. As described in Section 2.4, this means that both the left side and the right side of this equation must be *constant*. You can set that constant (called the "separation constant") equal to α and write

$$\frac{1}{X}\frac{\partial^2 X}{\partial x^2} = \alpha,$$
$$\frac{1}{v^2}\frac{1}{T}\frac{\partial^2 T}{\partial t^2} = \alpha$$

or

$$\frac{\partial^2 X}{\partial x^2} = \alpha X, \tag{3.20}$$
$$\frac{\partial^2 T}{\partial t^2} = \alpha v^2 T. \tag{3.21}$$

Now think about the meaning of these equations. They say that the second derivative of each function (X and T) is equal to a constant times that function. What kind of function will satisfy that requirement? As discussed in Chapter 2, harmonic functions (sines and cosines) fit this bill nicely.

To determine the arguments of the harmonic functions that satisfy Eqs. (3.20) and (3.21), set the constant α equal to $-k^2$. To understand why it's beneficial to make this constant negative and squared, consider the equation for $X(x)$:

$$\frac{\partial^2 X}{\partial x^2} = \alpha X = -k^2 X. \tag{3.22}$$

Solutions to this equation include $\sin(kx)$ and $\cos(kx)$, as you can verify by substituting $\sin(kx)$ or $\cos(kx)$ into Eq. (3.22). And when kx appears as the argument of a sine or cosine function, the meaning of k becomes clear: Since x represents distance, k must convert the dimension of distance to angle (in SI units, meters to radians). As described in Chapter 1, this is precisely the role of wavenumber ($k = 2\pi/\lambda$). So, by setting the separation constant α to $-k^2$, you ensure the explicit appearance of the wavenumber in your solution to the wave equation.

What about the equation for $T(t)$? In that case, setting the separation constant equal to $-k^2$ gives

$$\frac{\partial^2 T}{\partial t^2} = \alpha v^2 T = -v^2 k^2 T, \tag{3.23}$$

for which the solutions include $\sin(kvt)$ and $\cos(kvt)$. This means that kv must represent angular frequency (ω), which is exactly the case if k represents wavenumber and v represents wave phase speed (since $(2\pi/\lambda)v = 2\pi\lambda f/\lambda = 2\pi f = \omega$).

So separating the variables in the wave equation leads to solutions in the form of $y(x,t) = X(x)T(t)$, where the $X(x)$ functions may be $\sin(kx)$ or $\cos(kx)$ and the $T(t)$ terms may be $\sin(kvt)$ or $\cos(kvt)$. That's useful, but how do you know whether to choose the sine or cosine (or some combination of them) for the spatial function $X(x)$ and for the time function $T(t)$?

The answer is boundary conditions. Knowing the values that the solution (or its derivatives) must have at specified locations in space or time allows you to select the appropriate functions (those that match the boundary conditions). Those conditions may be points of zero displacement at the ends of a string, conducting planes near the source of electromagnetic waves, or potential barriers in the case of quantum waves.

Before working through an example showing how boundary conditions are applied, it's worth taking a few minutes to make sure you understand why the combination of sines and cosines (or even better, the *weighted* combination of sines and cosines) provides a more-general solution than either sines or cosines. To see that, consider what happens to the functions $X(x) = \sin(kx)$ or $X(x) = \cos(kx)$ at $x = 0$. The function $X(x) = \sin(kx)$ must equal zero at x = 0 (since $\sin 0 = 0$) , and the function $X(x) = \cos(kx)$ must equal its maximum value at x = 0 (since $\cos 0 = 1$). But what if your boundary conditions are such that the displacement is neither zero nor its maximum value at $x = 0$? If you limit yourself to using only sine functions or only cosine functions, your solution can never satisfy the boundary conditions.

Now consider what happens if you define your function $X(x)$ as a weighted combination of sine and cosine functions, so $X(x) = A\cos(kx) + B\sin(kx)$. The weighting coefficients A and B tell you "how much cosine" and "how much sine" to mix into the combination, and adding in just the right amount of each can have a very powerful effect.

To see that, take a look at the four waveforms in Fig. 3.4. Each of these waveforms represents $X(x) = A\cos(kx) + B\sin(kx)$, but the relative values of the weighting factors A and B have been varied. As you can see on the graph, the waveform peaking at $kx = 0$ (farthest to the left) has $A = 1$ and $B = 0$, so this is a pure cosine function. The waveform peaking at $kx = 90°$ (farthest to the right) has $A = 0$ and $B = 1$, which means it's a pure sine function.

Now look at the two waveforms that peak in between the pure cosine and the pure sine function. The waveform with $A = 0.866$ and $B = 0.5$ (labeled

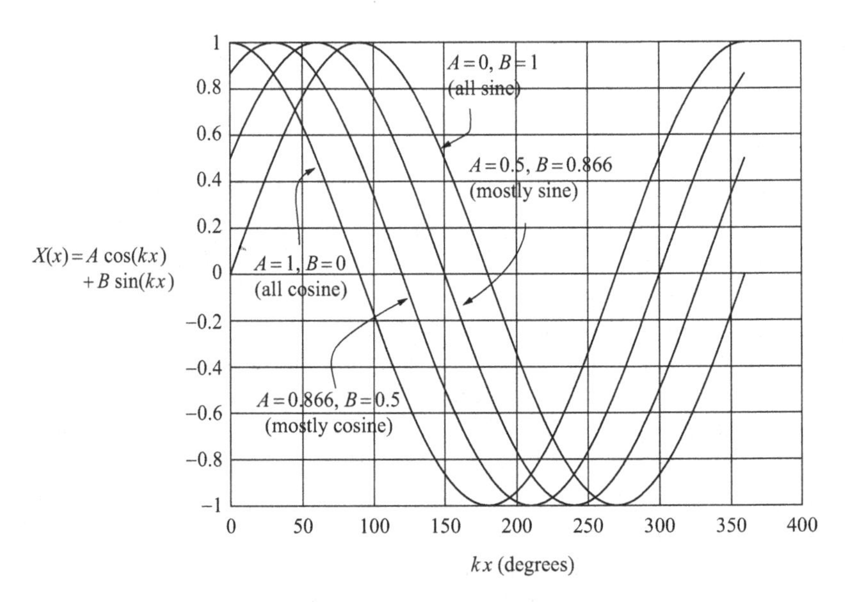

Figure 3.4 Weighted combinations of sine and cosine functions.

"mostly cosine") peaks at $kx = 30°$, and the waveform with $A = 0.5$ and $B = 0.866$ (labeled "mostly sine") peaks at $kx = 60°$. As you can see, the proportion of sine and cosine functions determines the left–right position of the peaks of the combined waveform. In this case, we've kept $A^2 + B^2 = 1$ to ensure that all of the waveforms have the same height, but if the boundary conditions call for a specific value of $X(x)$ at certain values of x, the weighting factors A and B can be adjusted to satisfy those conditions.

This explains why you so often encounter statements such as "The most general solution is a combination of these functions" when reading about waves. By combining functions such as sines and cosines with appropriate weighting factors, your solutions to the wave equation can satisfy a far greater range of conditions than can be achieved by sines or cosines alone.

The very useful ability to shift wavefunctions along the x- or t-axes can be achieved in other ways, such as using a phase constant (ϕ_0) and writing the general solution as $C\sin(\omega t + \phi_0)$. In that case, the height of the waveform is matched to the boundary conditions by adjusting the constant C, while the position of the peaks along the time axis is controlled by the phase constant ϕ_0.

The following example illustrates the application of boundary conditions to the problem of a string that is constrained at specified locations.

Example 3.3 *Find the displacement y(x,t) produced by waves on a string fixed at both ends.*

Since the string is fixed at both ends, you know that the displacement $y(x, t)$ must be zero for all time at the locations corresponding to the ends of the string. If you define one end of the string to have value $x = 0$ and the other end to have value $x = L$ (where L is the length of the string), you know that $y(0, t) = 0$ and $y(L, t) = 0$. Separating $y(x, t)$ into the product of distance function $X(x) = A\cos(kx) + B\sin(kx)$ and time function $T(t)$ means that

$$\begin{aligned} y(0, t) = X(0)T(t) = [A\cos(0) + B\sin(0)]\,T(t) &= 0, \\ [(A)(1) + (B)(0)]\,T(t) &= 0. \end{aligned}$$

Since this must be true at all time (t), this means that the weighting coefficient A for the cosine term must equal zero. Applying the boundary condition at the other end of the string ($x = L$) is also useful:

$$\begin{aligned} y(L, t) = X(L)T(t) = [A\cos(kL) + B\sin(kL)]\,T(t) &= 0, \\ [0\cos(kL) + B\sin(kL)]\,T(t) &= 0. \end{aligned}$$

Once again invoking the fact that this must be true over all time, this can only mean that either $B = 0$ or $\sin(kL) = 0$. Since $B = 0$ corresponds to the supremely boring case of no displacement anywhere on the string at any time (remember that you already know that $A = 0$), you'll have more fun if you consider the case for which B is non-zero and $\sin(kL)$ is zero. You know that $k = 2\pi/\lambda$, so in this case

$$\begin{aligned} \sin(kL) = \sin\left(\frac{2\pi L}{\lambda}\right) &= 0, \\ \frac{2\pi L}{\lambda} &= n\pi, \\ \lambda &= \frac{2L}{n}, \end{aligned}$$

where n can be any positive integer (taking n to be zero or negative doesn't lead to any interesting physics).

This means that for a string fixed at both ends, the wavelength (λ) takes on values of $2L$ (if $n = 1$), L (if $n = 2$), $2L/3$ (if $n = 3$), and so on. You can see what each of these cases looks like in Fig. 3.5.

So $\lambda = 2L/n$ with positive integer n defines the allowed wavelengths for a string fixed at both ends. Any value of λ between these values will not have zero displacement at the ends of the string and won't satisfy the boundary

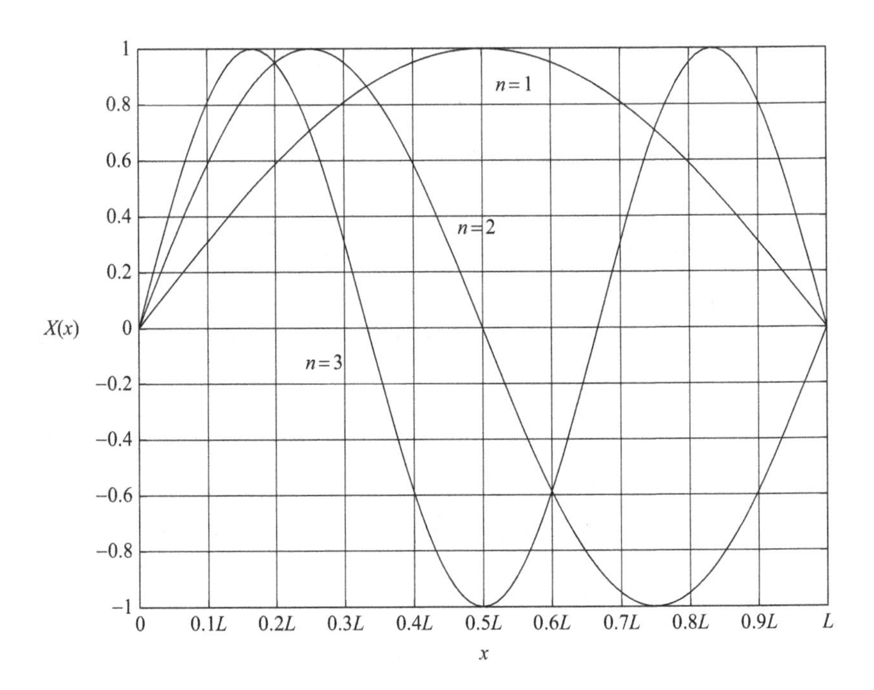

Figure 3.5 The first three modes for string fixed at both ends.

conditions of $X(0) = 0$ and $X(L) = 0$. So the general solution will be some weighted combination of these waveforms:

$$X(x) = B_1 \sin\left(\frac{\pi x}{L}\right) + B_2 \sin\left(\frac{2\pi x}{L}\right) + B_3 \sin\left(\frac{3\pi x}{L}\right) + \cdots$$

$$= \sum_{n=1}^{\infty} B_n \sin\left(\frac{n\pi x}{L}\right),$$

where the weighting coefficients B_n determine exactly how much of each of these waveforms is present. And the size of those coefficients depends on how you excite the string.

For example, you could "pluck" the string by pulling the string to some initial displacement at one or more locations. In that case, one boundary condition would specify the amount of initial displacement $y(x, 0)$ at each location at time $t = 0$, and another boundary condition would specify that the initial transverse velocity $\partial y(x, t)/\partial t$ equals zero at time $t = 0$.

Alternatively, you could "strike" the string to give it an initial transverse velocity at a specified location. In that case, one boundary condition would specify that the initial displacement $y(x, 0) = 0$ at each location at time

$t = 0$, and another boundary condition would specify that the initial transverse velocity $\partial y(x, t)/\partial t$ equals the imparted velocity v_0 at time $t = 0$.

To see how that works, imagine that the string is initially at equilibrium (no displacement, so $y(x, 0) = 0$ for all x) when you strike the string with a small hammer, imparting initial transverse velocity v_0. Separating the space and time components as described above and writing $T(t)$ as a weighted combination of sines and cosines gives

$$T(t) = C\cos(kvt) + D\sin(kvt) = C\cos\left(\frac{2\pi}{\lambda}vt\right) + D\sin\left(\frac{2\pi}{\lambda}vt\right),$$

in which the weighting coefficients C and D are determined by the boundary conditions.

The analysis of $X(x)$ has told you that $\lambda = 2L/n$ for a string fixed at both ends, so the expression for $T(t)$ becomes

$$\begin{aligned} T(t) &= C\cos\left(\frac{2\pi}{\lambda}vt\right) + D\sin\left(\frac{2\pi}{\lambda}vt\right) \\ &= C\cos\left(\frac{2\pi}{2L/n}vt\right) + D\sin\left(\frac{2\pi}{2L/n}vt\right) \\ &= C\cos\left(\frac{n\pi}{L}vt\right) + D\sin\left(\frac{n\pi}{L}vt\right). \end{aligned}$$

With this expression in hand, you can apply the boundary condition of zero displacement at time $t = 0$ for all values of x:

$$\begin{aligned} y(x, 0) = X(x)T(0) = X(x)[C\cos(0) + D\sin(0)] &= 0, \\ X(x)[(C)(1) + (D)(0)] &= 0. \end{aligned}$$

Since you know that $X(x) = \sum_{n=1}^{\infty} B_n \sin(n\pi x/L)$, which is not zero for all values of x, this means that C must equal zero. Thus the general solution for $T(t)$ is the sum of the sine terms for each value of n:

$$T(t) = \sum_{n=1}^{\infty} D_n \sin\left(\frac{n\pi}{L}vt\right).$$

Combining the expressions for $X(x)$ and $T(t)$ and absorbing the D_n weighting coefficients into B_n makes the solution for displacement

$$y(x, t) = X(x)T(t) = \sum_{n=1}^{\infty} B_n \sin\left(\frac{n\pi x}{L}\right) \sin\left(\frac{n\pi vt}{L}\right). \tag{3.24}$$

The $n = 1$ term of this expression is plotted at 50 different times between $kvt = 0$ and $kvt = 2\pi$ in Fig. 3.6. As you can see in this figure, this is an example of a standing wave. For this $n = 1$ case, there are two nodes

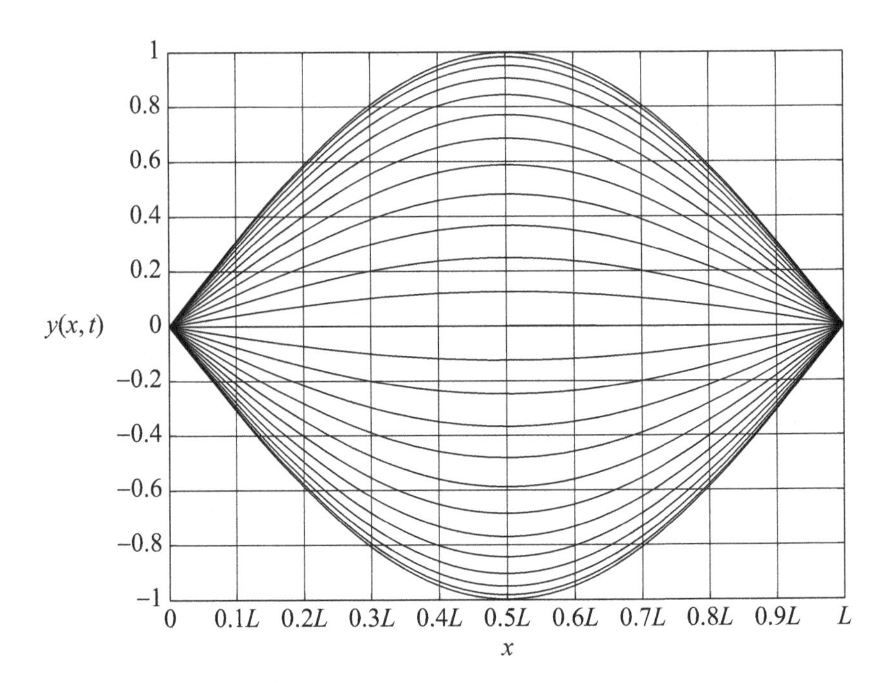

Figure 3.6 $y(x, t)$ for $n = 1$ at 50 times between $kvt = 0$ and 2π.

(locations of zero displacement), one at each end of the string, and one antinode (location of maximum displacement) at the center. You can relate the shape of this plot to the equation for $y(x, t)$ (Eq. (3.24)) by considering the role of each of the two sine terms. The spatial sine term, which is $\sin(n\pi x/L)$ (or $\sin(\pi x/L)$ for $n = 1$), produces one half-cycle of a sine wave over the distance $x = 0$ to $x = L$ (since the argument of $\sin(\pi x/L)$ advances from 0 to π over this range of x).

To understand the effect of the temporal sine term, $\sin(n\pi vt/L)$ (or $\sin(\pi vt/L)$ for $n = 1$), recall that $vt = (\lambda f)t$, and that $f = 1/T$, where T represents the period of oscillation. Thus $n\pi vt/L = n\pi\lambda t/(TL)$, and, since $\lambda = 2L/n$ for a string fixed at both ends, this becomes $2n\pi Lt/(nTL) = 2\pi t/T$. In this form, it's easier to see that this term varies from a value of zero at $t = 0$ (since $\sin 0 = 0$) to a maximum of +1 at time $t = T/4$ (since $\sin[2\pi(T/4)/T] = \sin(\pi/2) = 1$), back to zero at time $t = T/2$ (since $\sin[2\pi(T/2)/T] = \sin(\pi) = 0$), to a negative maximum of -1 at time $t = 3T/4$ (since $\sin(2\pi(3T/4)/T) = \sin(3\pi/2)) = -1$, and back to zero at time $t = T$ (since $\sin(2\pi(T)/T) = \sin(2\pi) = 0$). It is these temporal oscillations that are visible as the lines "filling in" the half-sine wave in Fig. 3.6.

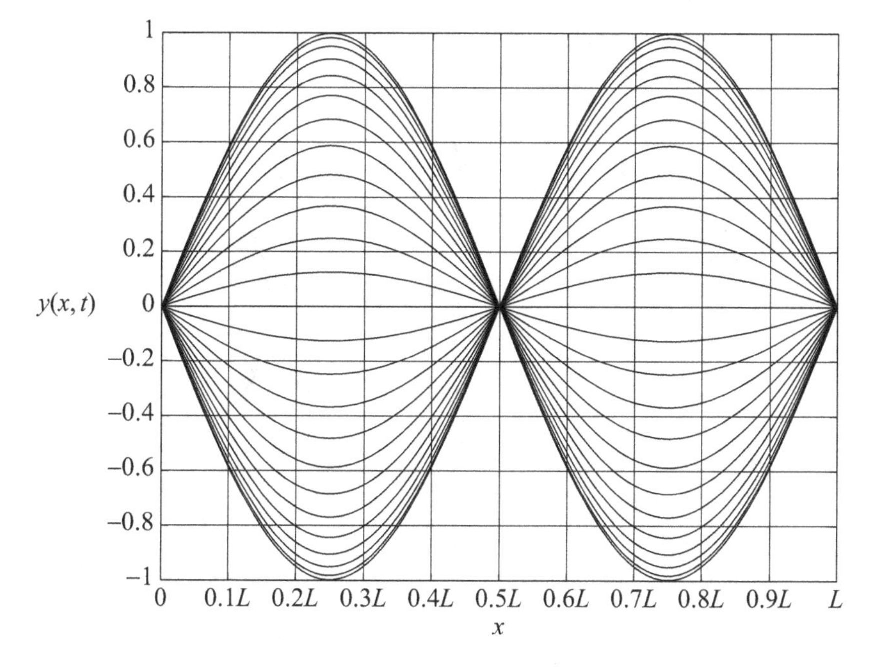

Figure 3.7 $y(x, t)$ for $n = 2$.

Similar analysis applies to the case of $n = 2$, except in this case the spatial sine term results in a full cycle over the distance from $x = 0$ to $x = L$, as you can see in Fig. 3.7. Notice that in this case there are three nodes and two anti-nodes; the general rule is that there will be $n + 1$ nodes and n anti-nodes for a string fixed at both ends. You can see this for the $n = 3$ case in Fig. 3.8, for which the spatial sine term results in 1.5 cycles over the distance from $x = 0$ to $x = L$.

These modes of oscillation of a string fixed at both ends are called "normal" modes, and the value of n is called the "order" of the mode. The mode with $n = 1$ is called the fundamental mode of vibration, and modes with larger values of n are called higher-order modes. You may also see normal modes described as "eigenfunctions", especially when dealing with the quantum waves described in Chapter 6.

The techniques described in this section have provided a good deal of information about the waveforms that may appear on a string, but you may have noticed that we haven't discussed exactly how the value of the weighting coefficients such as B_n in Eq. (3.24) can be determined from the boundary conditions. To understand that process, you need to be familiar with the basics of Fourier theory, which is the subject of the next section.

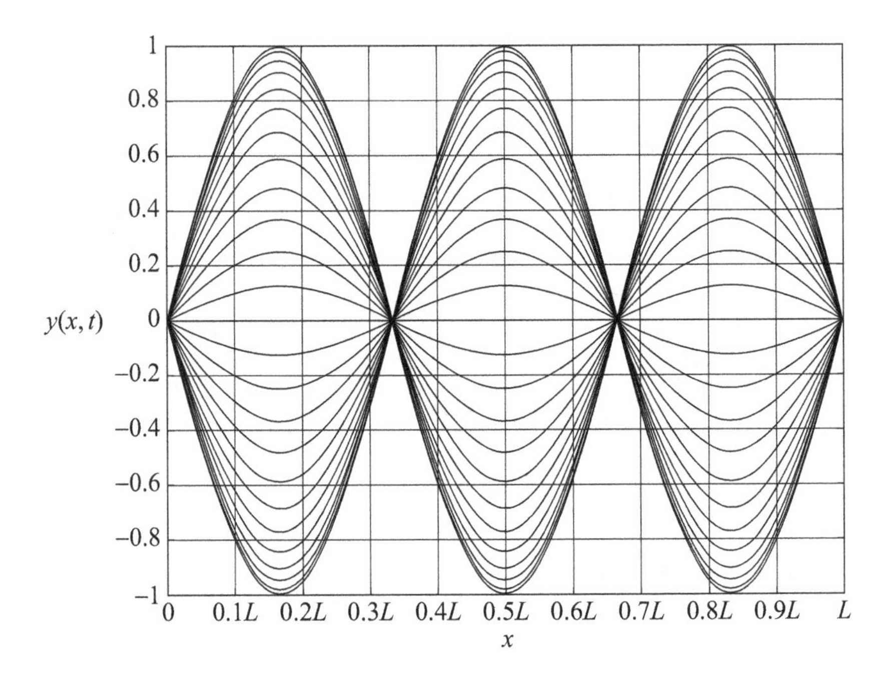

Figure 3.8 $y(x, t)$ for $n = 3$.

3.3 Fourier theory

As described in the introduction to this chapter, Fourier theory encompasses two related but distinct topics: Fourier synthesis and Fourier analysis. As its name implies, the subject of Fourier synthesis is the synthesis of a resultant waveform by putting together the right mix of frequency components using sines and cosines (these are called the "basis functions"). In Fourier analysis, the goal is to "deconstruct" a given waveform into the frequency components that make up that waveform and to find the amplitude and phase of those components.

If you followed the developments of the previous two sections, you've already seen some examples of one of the important principles behind Fourier theory. That principle is called "superposition", which says that the sum of any solutions to the wave equation is also a solution. As described in Chapter 1, this is true because the wave equation is linear, since taking a second derivative is a linear operation. So d'Alembert's general solution to the wave equation (which represents a wave as the combination of two counter-propagating waves) and the use of a weighted combination of sines and cosines are both examples of superposition.

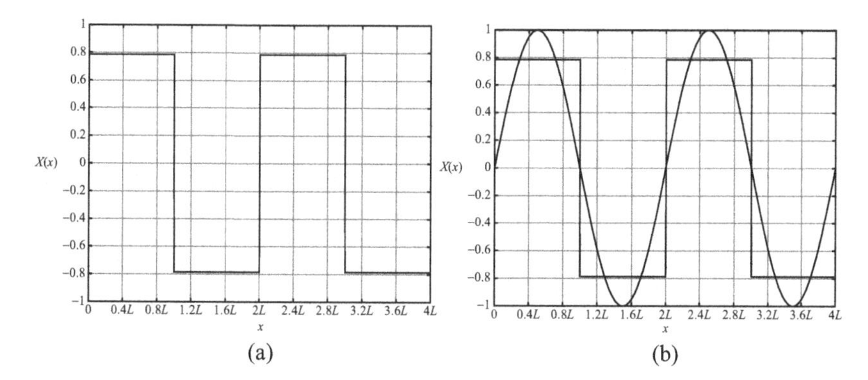

Figure 3.9 A square wave.

The real power of superposition was realized in the early 1800s when French mathematical physicist Jean Baptiste Joseph Fourier showed that even complex functions can be produced by combining a series of harmonic functions (sines and cosines). Fourier was studying heat flow along a thermally conducting material, but over the past 200 years his theory has been applied to many different areas of science and engineering.

As an example of how superposition works in Fourier synthesis, consider the square wave shown in Fig. 3.9(a). This figure shows two cycles of a wave that extends over all values of x from $-\infty$ to $+\infty$ and has a spatial period (the distance at which it repeats itself) of $2L$. It may seem implausible that this function, with its straight lines and sharp corners, can be constructed from the smoothly curving contours of sine and cosine functions. But it turns out that a series of sine waves with just the right amplitudes and frequencies can be made to converge to even this very straight-edged function.

To see that process in operation, take a look at the sine wave in Fig. 3.9(b). The spatial period of this wave (the distance of one full cycle) has been selected to match that of the square wave, as you can see from the fact that the zero crossings coincide with those of the square wave. The amplitude of the sine wave is about 27% bigger than the amplitude of the square wave; the reason for that will become clear when additional sine waves are added to this "fundamental" wave.

Although the frequency of the sine wave in Fig. 3.9(b) matches the frequency of the square wave in Fig. 3.9(a), the "square shoulders" and vertical jumps of the square wave are missing from the sine wave. But Figs. 3.10(a) and (b) show what happens when you add a second sine wave with frequency that is three times higher and amplitude that is 1/3 of the fundamental. Figure 3.10(a) shows that sine wave, and you can see that, at the locations at which the

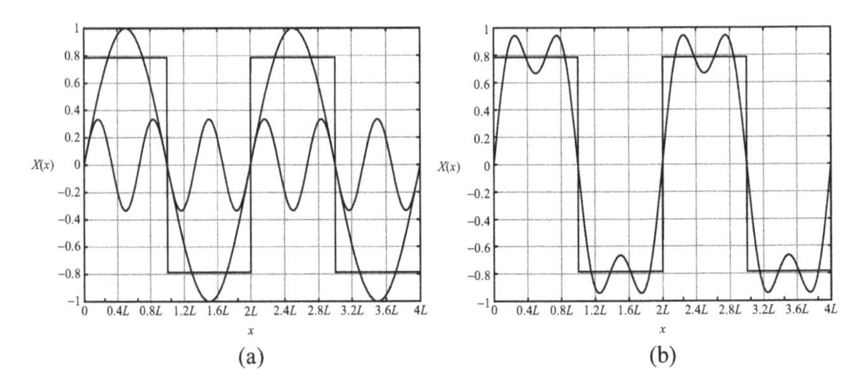

Figure 3.10 Two sine terms (a) and their sum (b).

fundamental overshoots the value of the square wave, this second wave has a negative value. So adding it to the fundamental wave reduces that overshoot and makes the value of the sum closer to the value of the square wave. Likewise, at the locations where the fundamental has values that are lower than the square wave, the second wave has positive values, so it tends to "fill in" the low points. If you look at the sum of the fundamental and the second sine wave in Fig. 3.10(b), the approximation to the square wave has gotten much better, although more can be done.

As you've probably guessed, the next step is to add in another sine wave, and this third component has frequency five times that of the fundamental and amplitude that is 1/5 that of the fundamental. Why those values of frequency and amplitude? As you can see in Fig. 3.11(a), this third sine wave has positive values at several locations at which the values of the sum of the fundamental and the second sine wave fall short of the square wave and negative values at locations at which they overshoot. Adding this third sine wave into the mix results in the combined waveform shown in Fig. 3.11(b), which is clearly a better approximation to the square wave.

If you continue to add more sine waves with the proper frequencies and amplitudes, the sum gets closer and closer to the desired square wave. In this case, the frequencies must be odd integer multiples of the fundamental frequency and the amplitudes must decrease by the same factors. So the terms of the series of sine waves that converge on this square wave are $\sin(\pi x/L)$, $\frac{1}{3}\sin(3\pi x/L)$, $\frac{1}{5}\sin(5\pi x/L)$ and so on. You can see the result of adding 16 terms in Fig. 3.12(a) and the result of adding 64 terms in Fig. 3.12(b). The approximation to the square wave gets better as you add more terms, and, although the series has an infinite number of terms, you don't need

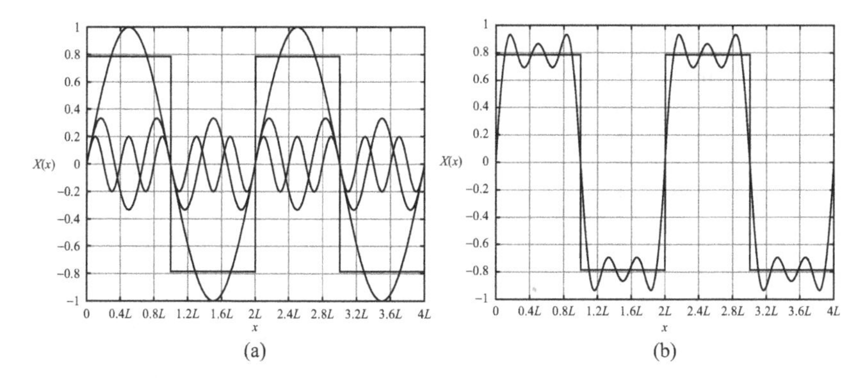

Figure 3.11 Three sine terms (a) and their sum (b).

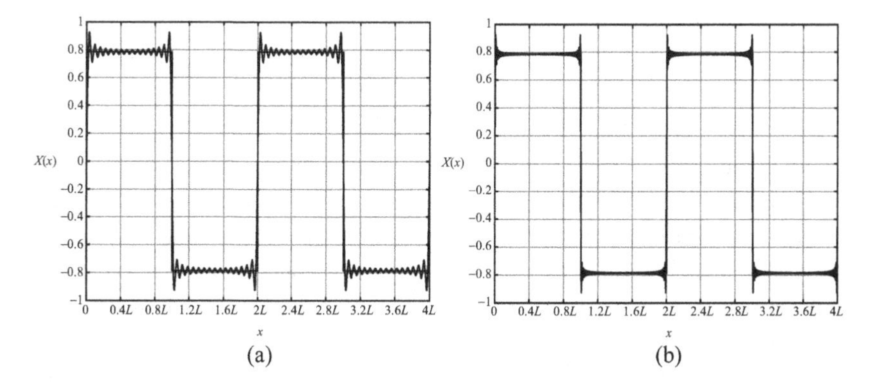

Figure 3.12 Sums of 16 sine terms (a) and 64 sine terms (b).

hundreds of terms to make the difference between the sum of the components and the ideal square wave reasonably small.

Among the differences that will always be with you are the small overshoots and oscillations just before and after the vertical jumps in the square waves. This is called "Gibbs ripple" and it will cause an overshoot of about 9% at the discontinuities of the square wave no matter how many terms of the series you add. But, as you can see in Figs. 3.12(a) and 3.12(b), adding more terms increases the frequency of the Gibbs ripple and reduces its horizontal extent in the vicinity of the jumps.

Although figures such as Figs. 3.10 and 3.11 are useful for illustrating the process of Fourier synthesis, there's a much more efficient way to display the frequency components of a wave. Instead of drawing the wave itself over space or time (that is, making a graph with the value of the displacement on the vertical axis and distance or time on the horizontal axis), a bar graph

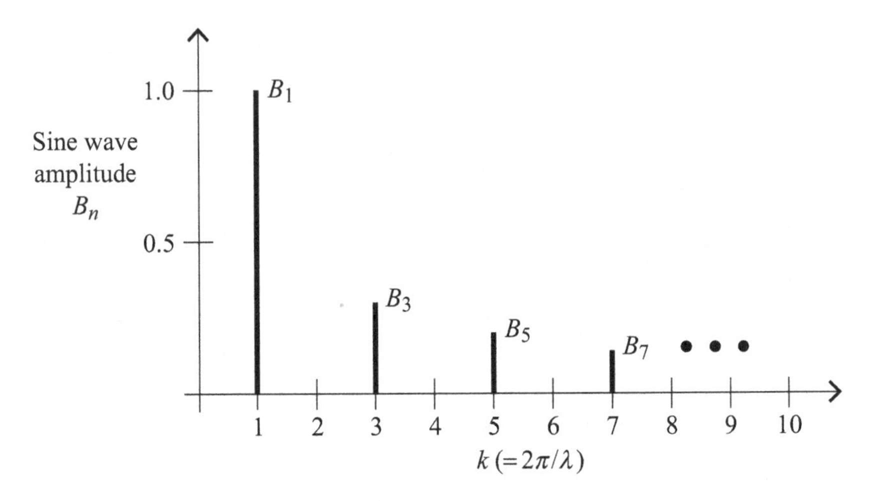

Figure 3.13 The single-sided sine-wave spectrum of a square wave.

showing the amplitude of each frequency component can be made. This type of graph is called a "frequency spectrum" and typically shows the amplitude of each frequency component on the vertical axis and frequency (f or ω) or wavenumber (k) on the horizontal axis. You can see an example of a wavenumber spectrum (also called a "spatial frequency domain plot") in Fig. 3.13.

This figure shows the amplitude of the first four sine waves that make up this square wave (these are B_1, B_3, B_5, and B_7; recall that the components for even values of n are all zero for this square wave). Had you wished your square wave to oscillate between the values of $+1$ and -1 (rather than $\pm\,0.785$ as in Fig. 3.9(a)), you could have multiplied the coefficient of each of these frequency components by a factor of $4/\pi \approx 1.27$.

This is an example of a "single-sided" magnitude spectrum in which the magnitudes of the frequency components are shown only for positive frequencies. This makes it easy to see which frequency components (and how much of each) are present in the resultant wave, but the only way you can tell that these amplitudes apply to sine waves (rather than cosine waves, for example) is the labeling on the graph. A full "two-sided" spectrum shows the amplitudes of the components for both positive and negative frequencies; such spectra contain all the information about the rotating phasors that make up sine and cosine waves (as described in Section 1.7 of Chapter 1). The two-sided spectrum of a single-frequency cosine wave is shown in Fig. 3.14(a), and the two-sided spectrum of a single-frequency sine wave is shown in

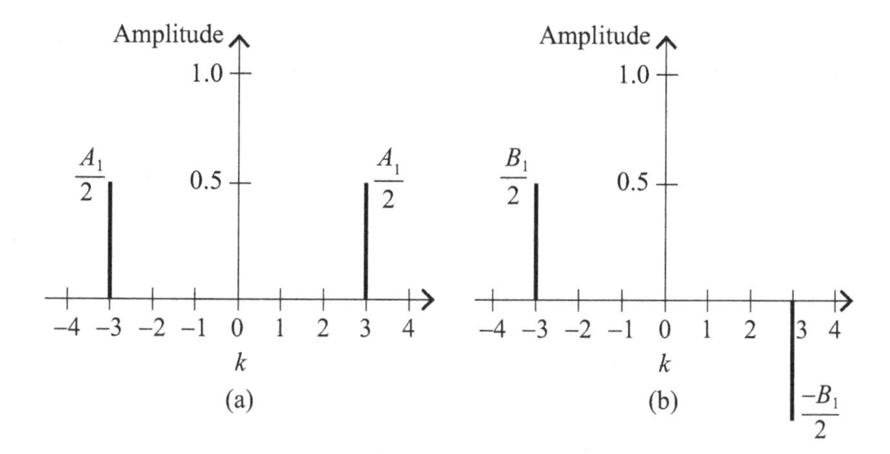

Figure 3.14 Two-sided spectra of a cosine wave (a) and a sine wave (b).

Fig. 3.14(b). Both of these waves have wavenumber $k = 3$. Notice that, for the full two-sided spectrum, the height of the bar on each side of zero frequency is *half* the value of the cosine (A_n) or sine (B_n) coefficients, which makes them consistent with the Euler equations for cosine (Eq. (1.43)) and sine (Eq. (1.44)). Notice also that both the negative-frequency component and the positive-frequency component of the cosine wave are positive, while the positive-frequency component of the sine wave is negative. This is consistent with the signs of the counter-rotating phasors in Eqs. (1.43) and (1.44).[3]

You're likely to encounter Fourier synthesis applied to both spatial wavefunctions such as $X(x)$ and time wavefunctions such as $T(t)$. In spatial applications, the resultant wavefunction is made up of spatial frequency components (also called wavenumber components) that are periodic (that is, they repeat themselves) over distance. The mathematical statement of Fourier synthesis for a spatial wavefunction with period $2L$ is

$$X(x) = A_0 + \sum_{n=1}^{\infty}\left[A_n \cos\left(\frac{n2\pi x}{2L}\right) + B_n \sin\left(\frac{n2\pi x}{2L}\right)\right]. \qquad (3.25)$$

In this expression, the A_0 term represents the constant (non-oscillating) average value (also called the "DC value" since a direct current does not oscillate) of $X(x)$. The A_n coefficients (A_1, A_2, A_3 etc.) tell you how much of each cosine component to add in and the B_n coefficients (B_1, B_2, B_3 etc.) tell you how much

[3] If you're wondering why the positive-frequency rather than the negative-frequency component is shown as negative, recall that the Euler equation for the sine function has a factor of i in the denominator, and $1/i = -i$.

of each sine component to add in to the mix to produce the function $X(x)$. We haven't cancelled the factors of 2 in the numerator and denominator of the sine and cosine arguments in Eq. (3.25) in order to make the role of the spatial period ($2L$) explicit.

In temporal (time-based) applications, the resultant wavefunction is made up of temporal frequency components that are periodic over time. The mathematical statement for a time function such as $T(t)$ with temporal period P is[4]

$$T(t) = A_0 + \sum_{n=1}^{\infty}\left[A_n \cos\left(\frac{n2\pi t}{P}\right) + B_n \sin\left(\frac{n2\pi t}{P}\right)\right], \tag{3.26}$$

in which the terms parallel those of Eq. (3.25), with the exception of the time period P replacing the spatial period $2L$ in the denominator of the sine and cosine arguments.

Consider again the Fourier coefficients for the DC term (A_0), the cosine components (A_n), and the sine component (B_n) that result in the square wave of Fig. 3.9(a). Changing the values of some or all of these coefficients can drastically change the nature of the function that results when all the terms of the series are added. For example, if you modify the square-wave coefficients by making the amplitude of the sine coefficients decrease as $1/n^2$ rather than as $1/n$ while also making every second coefficient negative, you get the waveform shown in Fig. 3.15. As you can see, these different coefficients result in a triangle wave rather than a square wave.

The equation for the sine coefficients (B_n) is shown to the right of the graph of the triangle wave; in this equation the $\sin(n\pi/2)$ term causes coefficients B_3, B_7, B_{11}, and so forth to be negative. The factor $8/\pi^2$ before each term causes the maximum and minimum values of the resultant wave to fall between $+1$ and -1.

Now take at look at the offset triangle wave shown in Fig. 3.16. Although this looks superficially like the triangle wave of Fig. 3.15, careful inspection reveals several important differences, suggesting that a different set of Fourier coefficients will be needed to synthesize this waveform.

One of those differences is that this triangle wave lies entirely above the x-axis, which means its average value is not zero. If you were tracking the previous discussion about the role of the Fourier coefficient called A_0, you might suspect that the "DC term" will not be zero in this case.

You can see another important difference in this wave by comparing the values on the left of $x = 0$ with those on the right. In this case, the values at

[4] We're using P to denote the period in order to avoid confusion with the function $T(t)$.

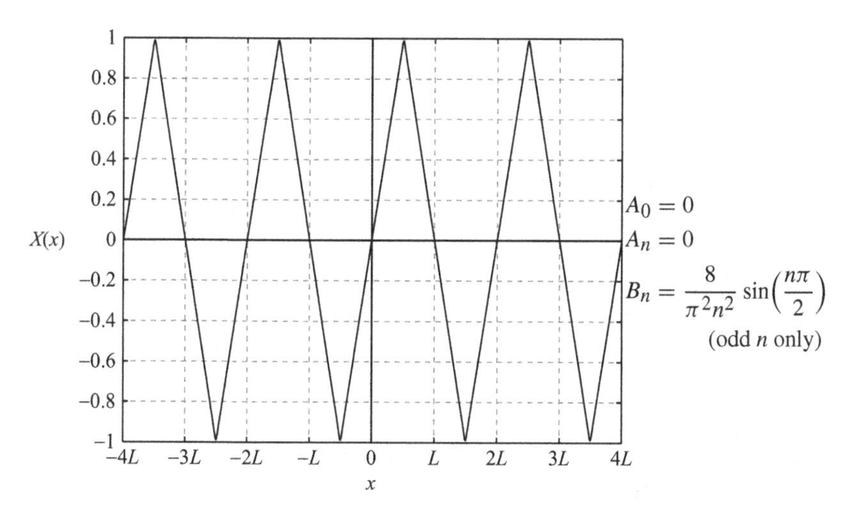

Figure 3.15 Fourier series for a periodic triangle wave.

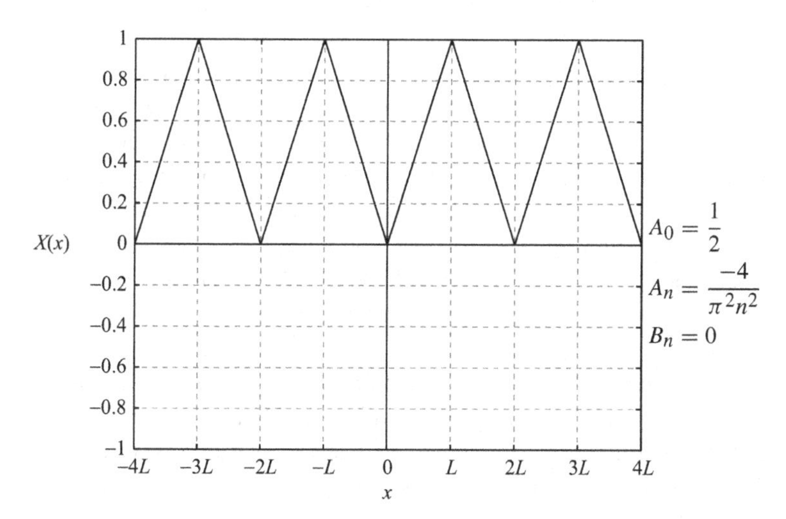

Figure 3.16 Fourier series for an offset periodic triangle wave.

equal distances on either side of $x = 0$ are equal (that is $X(-x) = X(x)$). That symmetry about $x = 0$ makes this an even function, whereas the triangle wave of Fig. 3.15 was an odd function, since in that case $X(-x) = -X(x)$. What does the evenness or oddness of a function tell you about the Fourier coefficients? Well, consider the nature of the sine function and the cosine function. Since the cosine function is an even function ($\cos(-x) = \cos(x)$) and the sine function is an odd function ($\sin(-x) = -\sin(x)$), any even function can be made up entirely of cosine components, and any odd function can be made up entirely of sine components.

Considering both the DC offset and the even nature of the triangle wave in Fig. 3.16, you can guess that the A_0 coefficient will be non-zero, the A_n (cosine) coefficients will also be non-zero, and the B_n (sine) coefficients will all be zero. And that guess is exactly right, as you can see in coefficient equations shown on the right side of Fig. 3.16.

At this point, you may be wondering whether *every* function can be represented by a Fourier series and exactly how the Fourier coefficients for a given function can be determined. The answer to the first question is provided by the Dirichlet requirements, and the answer to the second is the subject of Fourier analysis, which you can read about later in this section.

The Dirichlet requirements[5] state that the Fourier series converges (that is, gets closer to a limiting value as more terms are included) for any periodic function as long as that function has a finite number of extrema (maxima and minima) and a finite number of finite discontinuities in any interval. At locations at which the function is continuous (away from discontinuities), the Fourier series converges to the value of the function, and at the location of a finite discontinuity (which is a non-infinite jump in value), the Fourier series converges to the average of the values on either side of the discontinuity. So in a square wave, for example, at the location at which the function's value jumps from a high value to a low value, the Fourier series converges to the midpoint between the high and low values. You can read a great deal more about the Dirichlet requirements in the literature, and you should rest assured that there are many useful functions in science and engineering for which the Fourier series converges.

As mentioned above, the process of combining weighted sine and cosine components to produce a resultant waveform is called Fourier synthesis, and the process of determining which components are present in a waveform is called Fourier analysis. The key to Fourier analysis is the orthogonality of sine and cosine functions. Used in this way, "orthogonality" is a generalization of the concept of perpendicularity, meaning uncorrelated in a very particular sense.

To understand orthogonal functions, consider two functions $X_1(x)$ and $X_2(x)$. Imagine subtracting the mean value (so that both functions are (vertically) centered on zero) and then multiplying the value of X_1 at every value of x by the value of X_2 at the same value of x. Then take the results of those point-by-point

[5] These are sometimes called the "Dirichlet conditions", but they are not the same as the Dirichlet boundary conditions discussed in Section 3.2.

multiplications and add them up (if you have discrete values of x) or integrate them (if you have a continuous function of x). The result of that summation or integration will be zero if the functions $X_1(x)$ and $X_2(x)$ are orthogonal.

There's a graphical example illustrating this process below, but first you should look at the mathematical statement of orthogonality for harmonic sine and cosine functions (integer n and m):

$$\frac{1}{2L}\int_{-L}^{L}\sin\left(\frac{n2\pi x}{2L}\right)\sin\left(\frac{m2\pi x}{2L}\right)dx = \begin{cases}\frac{1}{2} \text{ if } n = m,\\ 0 \text{ if } n \neq m,\end{cases} \tag{3.27}$$

$$\frac{1}{2L}\int_{-L}^{L}\cos\left(\frac{n2\pi x}{2L}\right)\cos\left(\frac{m2\pi x}{2L}\right)dx = \begin{cases}\frac{1}{2} \text{ if } n = m > 0,\\ 0 \text{ if } n \neq m,\end{cases} \tag{3.28}$$

$$\frac{1}{2L}\int_{-L}^{L}\sin\left(\frac{n2\pi x}{2L}\right)\cos\left(\frac{m2\pi x}{2L}\right)dx = 0. \tag{3.29}$$

The integrations in these equations are performed over one complete cycle of the fundamental sine or cosine function (with spatial period of $2L$). The first of these equations tells you that two harmonic sine waves of different frequencies ($n \neq m$) are orthogonal to one another, while two sine waves of the same frequency ($n = m$) are non-orthogonal. Likewise, the second equation says that two harmonic cosine waves of different frequencies ($n \neq m$) are orthogonal, while two cosine waves of the same frequency ($n = m$) are non-orthogonal. The third equation tells you that harmonic sine and cosine waves are orthogonal irrespective of whether they have the same or different frequencies.

These orthogonality relations provide the perfect tool for determining which frequency components are present in a given waveform and which are not. To see how that works, imagine that you have a function that you wish to test for the presence of a certain frequency of sine wave. To conduct such a test, multiply the function you wish to test point-by-point by a "testing" sine wave and add up the results of those multiplications. If the function under test contains a frequency component that is a sine wave at the same frequency as the testing sine wave, each of the point-by-point multiplications will produce a positive result, and the sum of those results will be a large value. This process is illustrated in Fig. 3.17.

This example may appear trivial, since the function being tested is clearly a single-frequency sine wave. So it hardly seems necessary to go through the process of multiplying the value of this function by the testing function at every value of x and then integrating the results. But imagine the case in which the function being tested contains other frequency components of

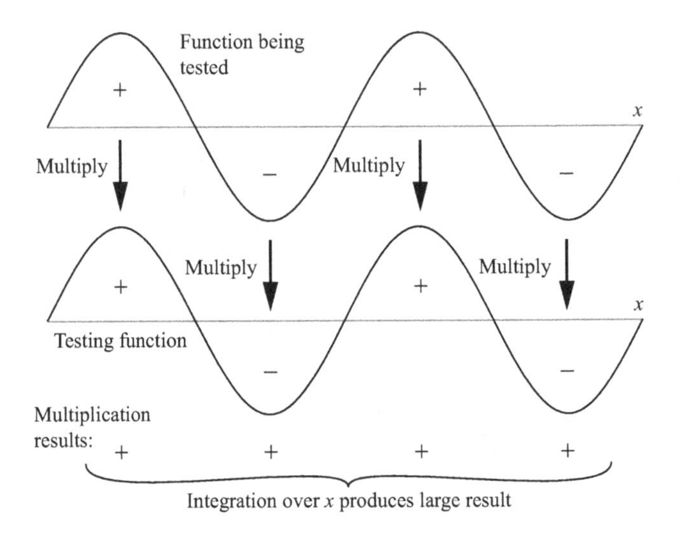

Figure 3.17 The frequency of the testing function matches a frequency present in the function being tested.

various amplitudes and frequencies in addition to the single sine wave shown in the figure. As long as the sine wave matching the frequency of the testing sine wave is present in the mix, the contribution to the multiplication and integration process from that frequency component will be positive (since all the positive portions of that sine wave will line up with all the positive portions of the testing function, and all the negative portions will line up as well). And the bigger the amplitude of the sine-wave component of the signal being tested that matches the frequency of the testing sine wave, the bigger the result of the multiplication and integration process will be. So this process doesn't just tell you that a certain frequency is present in the signal being tested, but gives you a result that is proportional to the amplitude of that frequency component.

When the frequency of the testing function does not match a frequency component of the signal being tested, the multiplication and integration process yields small results. To see an example of that, consider the case shown in Fig. 3.18.

In this case, the frequency of the testing sine wave is half the frequency of the sine wave in the function being tested. Now when you do point-by-point multiplication, the frequency component in the function being tested goes through a complete cycle over the distance through which the testing function goes through half a cycle. This means that some of the results of the multiplication are positive and some are negative, and, when you integrate across a complete cycle, the value is zero. So, in this case, the functions are orthogonal, as you would expect from Eq. (3.27) when $n \neq m$.

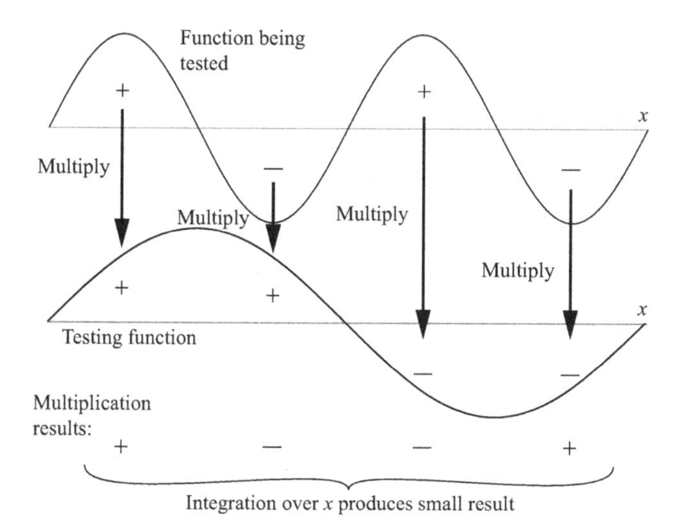

Figure 3.18 The frequency of the testing function is half the frequency of the function being tested.

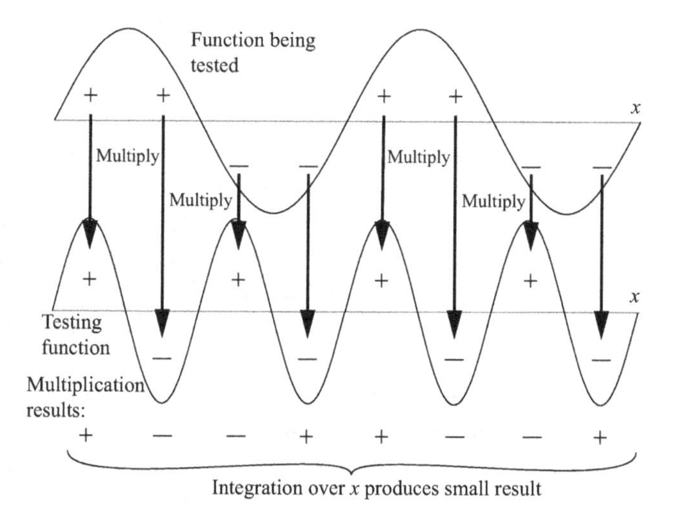

Figure 3.19 The frequency of the testing function is twice the frequency of the function being tested.

As you can see in Fig. 3.19, the same thing occurs when the frequency of the testing function is twice that of the frequency component in the function under test. In this case, the frequency component in the function being tested goes through only half a cycle in the distance over which the testing function goes through a complete cycle. As in the previous case, this means that some of the results of the multiplication are positive and some are negative, and, when you integrate across a complete cycle, the value is again zero. So, whenever the frequency component in the function being tested goes through more or fewer complete cycles than the testing function (in other words, whenever $n > m$ or $n < m$), the result of the multiplication and integration process will be zero.

If you think about the point-by-point multiplication process illustrated in Figs. 3.17 through 3.19, you may realize that, although the results are all shown as big "plus" and "minus" signs, the size of the result will vary depending on the value of x at which the multiplication is occurring. For example, in Fig. 3.17, the results will all be positive, but some of the multiplications will yield bigger products than others. In fact, the result of the multiplication will vary sinusoidally along the x-axis. So what you're really doing when you integrate the results of the multiplications is finding the area under the sinusoidal curve formed by the product of the function being tested and the testing function. That sinusoidal curve will lie entirely above the x-axis (all values positive) when the frequency of the testing function matches the frequency of the function being tested (that is, when $n = m$ in Eq. (3.27)), but it will lie half above and half below the x-axis (with as many negative values as positive ones) when the functions are orthogonal (when $n \neq m$ in Eq. (3.27)). And when a sinusoid lies entirely above the x-axis, the area under the curve is positive, but when a sinusoid lies half above and half below the x-axis, the area under the curve is zero. This is why the orthogonality relations work.

As you've probably guessed, the same process works for determining whether a certain cosine wave is present in the function being tested, although in that case the testing functions are cosine waves rather than sine waves. So you can find the Fourier A_n coefficients for any function satisfying the Dirichlet requirements by multiplying that function by cosine waves and integrating the results just as you can find the Fourier B_n coefficient by using the same process with sine waves. And the orthogonality of sine and cosine functions (Eq. (3.29)) guarantees that cosine components in the function being tested will contribute to the A_n coefficients but will add nothing to the B_n coefficients.

Using both sine waves and cosine waves as testing functions resolves a question that many students have about the multiplication and integration process illustrated above. Their concern is that the process works well when the sine-wave component in the function being tested is exactly "lined up" with the testing-function sine wave (that is, there's no phase offset between the frequency component and the testing sine wave). But here's the beauty of Fourier's approach: If the function being tested contains a component that has a certain frequency but is offset in phase (so that it "lines up" with neither the testing sine wave nor the testing cosine wave), the multiply-and-integrate process will yield a non-zero value both for the A_n coefficient and for the B_n coefficient at that frequency. So, if the frequency component in the function being tested is close to being in phase with the testing sine wave, the value of B_n will be large and the value of A_n will be small, but if the frequency component is close to being in phase with the testing cosine wave, the value of A_n will be large and the value of B_n will be small. If the phase of the frequency component is exactly in the middle between the phase of the testing sine wave and the phase of the testing cosine wave, then A_n and B_n will have equal value. This is an application of the superposition concept discussed earlier in this section and illustrated in Fig. 3.4.

Here are the mathematical statements of the processes by which you can find the values of the Fourier coefficients of a waveform $X(x)$:

$$\begin{aligned} A_0 &= \frac{1}{2L}\int_{-L}^{L} X(x)dx, \\ A_n &= \frac{1}{L}\int_{-L}^{L} X(x)\cos\left(\frac{n2\pi x}{2L}\right)dx, \\ B_n &= \frac{1}{L}\int_{-L}^{L} X(x)\sin\left(\frac{n2\pi x}{2L}\right)dx. \end{aligned} \tag{3.30}$$

Notice that to find the value of the non-oscillating component (the "DC" term) of $X(x)$ you just integrate the function; the "testing function" in this case has a constant value of one and the integration yields the average value of $X(x)$.

The following example may help you see how to use these equations.

Example 3.4 *Verify the Fourier coefficients shown for the triangle wave in Fig. 3.16. Assume that the spatial period (2L) is 1 meter, and the units of $X(x)$ are also meters.*

If you followed the discussion about this triangle wave, you already know that the DC term (A_0) and the cosine coefficients (A_n) should be non-zero and that the sine coefficients (B_n) should all be zero (since this wave is an even function with non-zero average value). You can verify those conclusions using Eqs. (3.30), but first you have to figure out the period of $X(x)$ and the equation for $X(x)$.

You can read the period right off the graph: This waveform repeats itself with a period of one meter. Since the spatial period is represented as $2L$ in the Fourier-series equations, this means that $L = 0.5$ meter. To determine the equation for $X(x)$, notice that this function is made up of straight lines, and the equation of a straight line is $y = mx + b$, where m is the slope of the line and b is the y-intercept (the value of y at the point at which the line crosses the y-axis).

You can choose to analyze any of the complete cycles shown on the graph, but in many cases you can save time and effort by selecting a cycle that's centered on $x = 0$ (you'll see why that's true later in this example). So instead of considering a cycle consisting of one of the triangles with the point at the top (such as the triangle between $x = 0$ and $x = 2L$), you can consider the "inverted triangle" (with the point at the bottom) between $x = -L$ and $x = L$.

The slope of the line between $x = -L = -0.5$ and $x = 0$ is -2 (because the "rise" is -1 and the "run" is 0.5, so the rise over the run is $-1/0.5 = -2$) and the y-intercept is zero. So the equation for this portion of $X(x)$ is $X(x) = mx + b = -2x + 0$. A similar analysis between $x = 0$ and $x = L = 0.5$ gives the equation $X(x) = mx + b = 2x + 0$. With these equations in hand, you can now plug $X(x)$ into the equation for A_0,

$$\begin{aligned} A_0 &= \frac{1}{2L}\int_{-L}^{L} X(x)dx = \frac{1}{2(0.5)}\left[\int_{-0.5}^{0} -2x\,dx + \int_{0}^{0.5} 2x\,dx\right] \\ &= (1)\left[-2\left(\frac{x^2}{2}\right)\Big|_{-0.5}^{0} + 2\left(\frac{x^2}{2}\right)\Big|_{0}^{0.5}\right] = 0 - (-0.25) + 0.25 - 0 \\ &= 0.5 \end{aligned}$$

and into the equation for A_n,

$$\begin{aligned} A_n &= \frac{1}{L}\int_{-L}^{L} X(x)\cos\left(\frac{n2\pi x}{2L}\right)dx \\ &= \frac{1}{0.5}\left[\int_{-0.5}^{0} -2x\cos(2n\pi x)dx + \int_{0}^{0.5} 2x\cos(2n\pi x)dx\right]. \end{aligned}$$

Using integration by parts (or looking up $\int x\cos(ax)dx$ in a table of integrals), you'll find that $\int x\cos(ax)dx = (x/a)\sin(ax) + (1/a^2)\cos(ax)$, so the equation for A_n becomes

$$\begin{aligned} A_n &= \frac{-2}{0.5}\left[\frac{x}{2n\pi}\sin(2n\pi x)|_{-0.5}^{0} + \frac{1}{4n^2\pi^2}\cos(2n\pi x)|_{-0.5}^{0}\right] \\ &\quad + \frac{2}{0.5}\left[\frac{x}{2n\pi}\sin(2n\pi x)|_{0}^{0.5} + \frac{1}{4n^2\pi^2}\cos(2n\pi x)|_{0}^{0.5}\right] \\ &= \frac{-2}{0.5}\left[0 - \frac{-0.5}{2n\pi}\sin(2n\pi(-0.5)) + \frac{1}{4n^2\pi^2}(1 - \cos(2n\pi(-0.5)))\right] \\ &\quad + \frac{2}{0.5}\left[\frac{0.5}{2n\pi}\sin(2n\pi(0.5)) - 0 + \frac{1}{4n^2\pi^2}(\cos(2n\pi(0.5)) - 1)\right]. \end{aligned}$$

Recall that $\sin(n\pi) = 0$ and $\cos(n\pi) = (-1)^n$, so

$$\begin{aligned} A_n &= \frac{-2}{0.5}\left[0 - 0 + \frac{1}{4n^2\pi^2}(1 - (-1)^n)\right] \\ &+ \frac{2}{0.5}\left[0 - 0 + \frac{1}{4n^2\pi^2}((-1)^n - 1)\right] \\ &= \frac{-4}{0.5}\left[\frac{1}{4n^2\pi^2}(1 - (-1)^n)\right] = \left[\frac{-2}{n^2\pi^2}(1 - (-1)^n)\right] \\ &= \frac{-4}{n^2\pi^2} \text{ for odd } n. \end{aligned}$$

Fortunately, determining the B_n coefficients for this waveform is much easier. Since

$$B_n = \frac{1}{L}\int_{-L}^{L} X(x)\sin\left(\frac{n2\pi x}{2L}\right)dx$$

you can see by inspection that B_n must be zero. What exactly is involved in that inspection? Well, you know that $X(x)$ is an even function, since it has the same values at $-x$ as it does at $+x$. You also know that the sine function is odd, since $\sin(-x) = -\sin(x)$, and the product of an even function (like $X(x)$) and an odd function (like the sine function) is odd. But when you integrate an odd function between limits that are symmetric about $x = 0$ (such as $\int_{-L}^{L}$), the result is zero. Hence you know that B_n must equal zero for all values of n. This is one reason why choosing the cycle between $x = -L$ and $x = L$ is advantageous in this case (another reason is that $\int_{-L}^{L}(\text{even function})dx = 2\int_0^L(\text{even function})dx$, and both $X(x)$ and the cosine function are even, so you could have simplified the calculation of the A_n coefficients as well).

So the Fourier coefficients for the triangle wave shown in Fig. 3.16 are indeed

$$A_0 = \frac{1}{2}, \qquad A_n = \frac{-4}{\pi^2 n^2}, \qquad B_n = 0$$

as expected from Fig. 3.16. If you'd like more practice at finding Fourier coefficients, you'll find additional problems like this at the end of this chapter, with full solutions on the book's website.

The next major topic of this section is the transition from discrete Fourier analysis of periodic waveforms to continuous Fourier transforms. But, before you try to understand that transition, it's worth taking the time to consider a different form of the equation for Fourier series (Eq. (3.25)). To see where this alternative form comes from, try expanding the sine and cosine functions in Eq. (3.25) into complex exponentials using the Euler relations from Chapter 1.

With a bit of algebraic manipulation, you can get to an alternative series for $X(x)$ that looks like this:

$$X(x) = \sum_{n=-\infty}^{\infty} C_n e^{i[n2\pi x/(2L)]}. \tag{3.31}$$

In this equation, the coefficients C_n are complex values produced by combining A_n and B_n. Specifically, $C_n = \frac{1}{2}(A_n \mp iB_n)$, and you can find the C_n coefficients directly from $X(x)$ using

$$C_n = \frac{1}{2L}\int_{-L}^{L} X(x)e^{-i[n2\pi x/(2L)]}\,dx. \tag{3.32}$$

There's no new physics in using the complex version of the Fourier series, but this form of the equations makes it considerably easier to see how the equation for the Fourier transform relates to Fourier series.

To understand that relationship, consider the difference between a periodic waveform (that is, a waveform that repeats itself after a finite interval of space or time) and a non-periodic waveform (a waveform that never repeats itself, no matter where or when you look). Periodic waveforms can be represented by discrete spatial frequency spectra (that is, spectra in which many wavenumbers have zero amplitude), such as the spectrum shown in Fig. 3.13.

The reason why you need only certain frequency components for a periodic waveform is illustrated in Fig. 3.20. This figure shows the first three spatial frequency components of a square wave, and the important thing to notice is that each of these sine-wave components *must* repeat itself at the same location as that at which the square wave repeats itself (although higher-frequency components may repeat themselves at earlier points). But any sine wave that

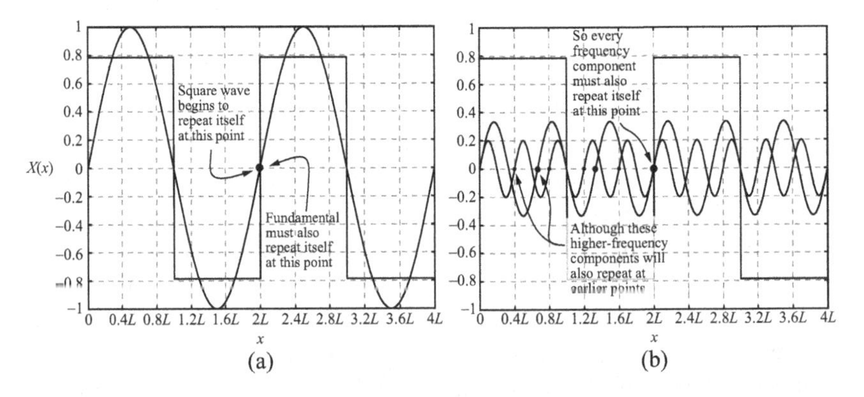

Figure 3.20 The periodicity of a square wave and component waves.

doesn't repeat itself at this point can't be a component of the square wave, because adding it to the mix that produces the square wave would cause the square wave not to repeat itself at this point.

This means that every wavenumber component of a periodic waveform must have an integer number of cycles in the interval over which the resultant waveform has one cycle. If the spatial period of the resultant waveform is $2L$, then the spatial frequency components that make up the waveform can only have wavelengths of $2L$, or $2L/2$, or $2L/3$, and so on. A wave with period $2L/1.5$ or $2L/3.2$ cannot be a frequency component of a waveform with period P. This is why the spectrum of a periodic waveform looks like a series of spikes rather than a continuous function.

You can see an example of this in the wavenumber spectrum of a train of rectangular pulses shown in Fig. 3.21. The envelope shape $K(k)$ of this spectrum is discussed below, but for now just observe the fact that there are many wavenumbers for which the spectral amplitude is zero. Stated another way, the spatial frequency components that make up this pulsetrain must repeat themselves in the same interval as that over which the pulsetrain repeats. Most wavenumbers don't do that, so their amplitude must be zero. So if the spatial period of the pulsetrain is $2L$, the spatial frequency components that do contribute to the pulsetrain must have wavelengths (λ) of $2L$, $2L/2$,

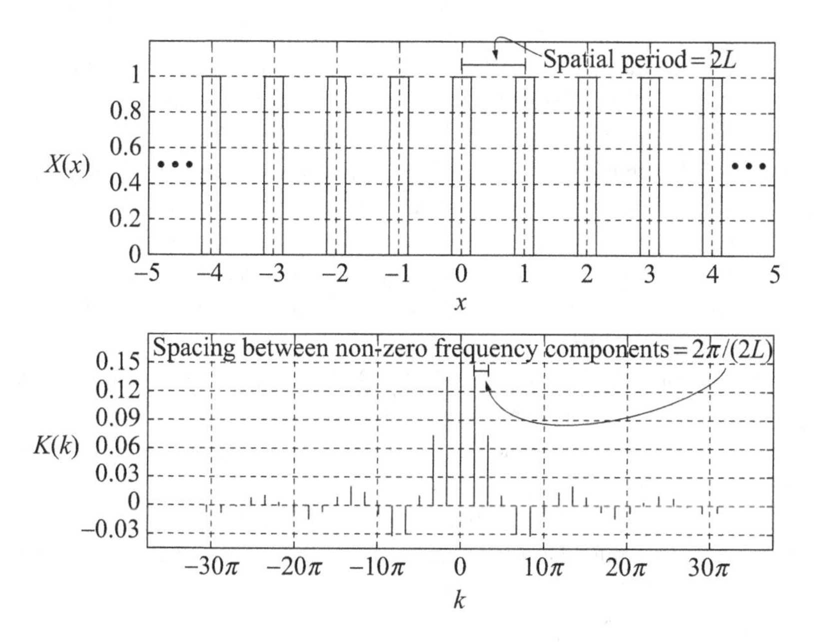

Figure 3.21 A periodic pulsetrain and its frequency spectrum.

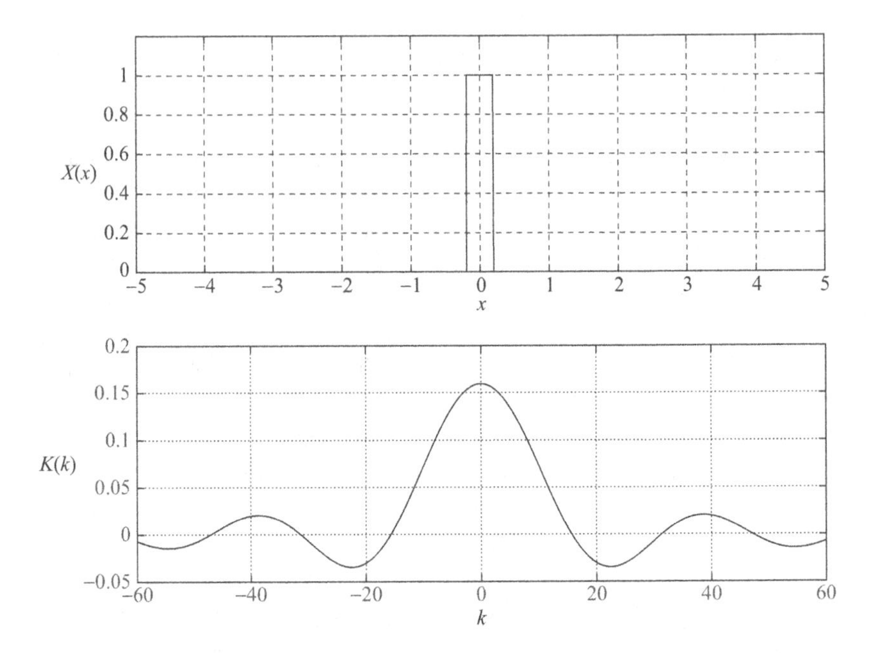

Figure 3.22 A single pulse (non-periodic) and its frequency spectrum.

$2L/3$, and so on, and, since wavenumber equals $2\pi/\lambda$, these spatial frequency components appear in the spectrum at wavenumbers of $2\pi/2L$, $4\pi/2L$, $6\pi/2L$, and so on. Thus the spacing between the wavenumber components is $2\pi/(2L)$.

Contrast the spectrum of the pulsetrain to the spectrum of a single, non-repeating pulse shown in Fig. 3.22. Since this waveform never repeats, its spectrum is a continuous function. One way to understand that is to consider its period P to be infinite, and to remember that the spacing of the frequency components that make up a waveform is proportional to $1/P$. Since $1/\infty = 0$, the frequency components of a non-periodic waveform are infinitely close together, and the spectrum is continuous.

The shape of the function $K(k)$ in Fig. 3.22 is extremely important in many applications in physics and engineering. It's called "sine x over x" or "sinc x", and it comes about through the continuous version of Fourier analysis called Fourier transformation.

Just as in the case of discrete Fourier analysis, the goal of Fourier transformation is to determine the wavenumber or frequency components of a given waveform such as $X(x)$ or $T(t)$. That process is closely related to the "multiply and integrate" process described above, as you can see in the mathematical statement of the Fourier transform:

$$K(k) = \frac{1}{\sqrt{2\pi}} \int_{-\infty}^{\infty} X(x) e^{-i(2\pi x/\lambda)}\, dx = \frac{1}{\sqrt{2\pi}} \int_{-\infty}^{\infty} X(x) e^{-ikx}\, dx, \qquad (3.33)$$

where $K(k)$ represents the function of wavenumber (spatial frequency) that is the spectrum of $X(x)$. The continuous function $K(k)$ is said to exist in the "spatial frequency domain" while $X(x)$ exists in the "distance domain". The amplitude of $K(k)$ is proportional to the relative amount of each spatial frequency component that contributes to $X(x)$, so $K(k)$ and $X(x)$ are said to be a Fourier transform pair. This relationship is often written as $X(x) \leftrightarrow K(k)$, and two functions related by the Fourier transform are sometimes called "conjugate variables".

Notice that there's no n in the continuous form of the Fourier transform; you're not restricted to using multiples of some fundamental frequency in this case.

If you have a spatial frequency-domain function such as $K(k)$, you can use the "inverse Fourier transform" to determine the corresponding distance-domain function $X(x)$. This process is the continuous equivalent to the discrete process of summing the terms of a Fourier series, and the inverse Fourier transform differs from the (forward) Fourier transform only by the sign of the exponent:

$$X(x) = \frac{1}{\sqrt{2\pi}} \int_{-\infty}^{\infty} K(k) e^{i(2\pi x/\lambda)}\, dk = \frac{1}{\sqrt{2\pi}} \int_{-\infty}^{\infty} K(k) e^{ikx}\, dk. \qquad (3.34)$$

For time-domain functions such as $T(t)$, there's an equivalent Fourier-transformation process which gives the frequency function $F(f)$:

$$F(f) = \int_{-\infty}^{\infty} T(t) e^{-i(2\pi t/T)}\, dt = \int_{-\infty}^{\infty} T(t) e^{-i(2\pi ft)}\, dt. \qquad (3.35)$$

Example 3.5 *Find the Fourier transform of a single rectangular distance-domain pulse $X(x)$ with height A over interval $2L$ centered on $x = 0$.*

Since the pulse is a distance-domain function, you can use Eq. (3.33) to transform $X(x)$ to $K(k)$. Since $X(x)$ has amplitude A between positions $x = -L$ and $x = L$ and zero amplitude at all other times, this becomes

$$\begin{aligned} K(k) &= \frac{1}{\sqrt{2\pi}} \int_{-\infty}^{\infty} X(x) e^{-ikx}\, dt = \frac{1}{\sqrt{2\pi}} \int_{-L}^{L} A e^{-ikx}\, dt \\ &= \frac{1}{\sqrt{2\pi}} A \frac{1}{-ik} e^{-ikx} \Big|_{-L}^{L} = \frac{1}{\sqrt{2\pi}} \frac{A}{-ik} \left[e^{-ikL} - e^{-ik(-L)} \right] \\ &= \frac{1}{\sqrt{2\pi}} \frac{2A}{k} \left[\frac{e^{-ikL} - e^{ikL}}{-2i} \right] = \frac{1}{\sqrt{2\pi}} \frac{2A}{k} \left[\frac{e^{ikL} - e^{-ikL}}{2i} \right]. \end{aligned}$$

But Euler says the term in square brackets is equal to $\sin(kL)$, so

$$K(k) = \frac{1}{\sqrt{2\pi}} \frac{2A}{k} \sin(kL)$$

and multiplying by L/L makes this

$$K(k) = \frac{A(2L)}{\sqrt{2\pi}} \left[\frac{\sin(kL)}{kL} \right].$$

This explains the $\sin(x)/x$ shape of the wavenumber spectrum of the rectangular pulse shown in Fig. 3.22.

There's an extremely important concept lurking in the Fourier transform pair of a rectangular pulse and the $\sin(x)/x$ function. You can see that concept in action by considering the wavenumber spectrum of the wider pulse shown in Fig. 3.23.

If you compare the spectrum of the wide pulse with that of the narrower pulse in Fig. 3.22, you'll see that the wider pulse has a *narrower* wavenumber spectrum $K(k)$ (that is, the width of the main lobe of the $\sin(x)/x$ function is smaller for a wide pulse than for a narrow pulse). And the wider you make the

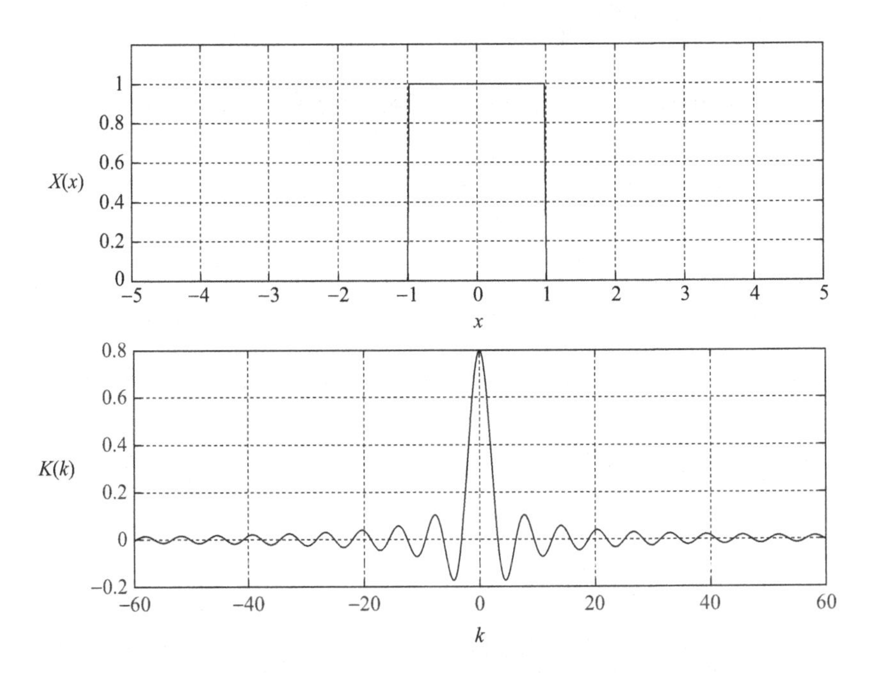

Figure 3.23 A wide pulse and its wavenumber spectrum.

pulse, the narrower the wavenumber spectrum becomes (it also gets taller, but it's the change in the width that contains some interesting physics).

That physics is called the "uncertainty principle", which you may have encountered if you've studied some modern physics (where it's usually called "Heisenberg's uncertainty principle"). But the uncertainty principle isn't restricted to the realm of quantum mechanics; it describes the relationship between any function and its Fourier transform, such as $X(x) \leftrightarrow K(k)$ or $T(t) \leftrightarrow F(f)$.

And what exactly does the uncertainty principle tell you? Just this: If a function is narrow in one domain, the Fourier transform of that function cannot also be narrow. Thus the frequency spectrum of a short time pulse cannot be narrow (which makes sense, since it takes lots of high-frequency components to make that pulse rise and then quickly fall). And if you have a very narrow spectrum (for example, a very sharp spike at a single frequency), the inverse Fourier transform gives you a function that extends over a large amount of distance or time.

This is why a true single-frequency signal has never been produced – it would have to continue for all time, with no starting or ending time. What if you were to take a small chunk out of a single-frequency sine or cosine wave? Then you'd have a signal with finite time extent, and, if that extent is Δt, then the frequency spectrum of that signal will have a width of about $1/\Delta t$. Take a very short sample of the wave (that is, make Δt very small), and you're guaranteed to have a very wide spectrum.

The mathematical statement of the uncertainty principle between the time domain and the frequency domain is

$$\Delta f \, \Delta t = 1,$$

where Δf represents the width of the frequency-domain function and Δt is the width of the time-domain function. The equivalent uncertainty principle for the distance/wavenumber domain is

$$\Delta x \, \Delta k = 2\pi,$$

where Δx represents the width of the distance-domain function and Δk is the width of the wavenumber-domain function.

These uncertainty relations tell you that you cannot know both time and frequency with high precision. If you make an extremely precise time measurement (so Δt is tiny), you cannot simultaneously know the frequency very accurately (since Δf will be large in order to make the product $\Delta f \Delta t$ equal to one). Likewise, if you know the location very precisely (so Δx is very small), you cannot simultaneously know the wavenumber very accurately (since Δk will be large). This principle will have important ramifications when it is applied to quantum waves in Chapter 6.

One last point about Fourier theory: Although Fourier used sines and cosines as his "basis functions" (that is, the functions from which other functions are produced), it's possible to use other orthogonal functions as basis functions (they must be orthogonal, or you couldn't find the coefficients using the techniques described above). Depending on the characteristics of the waveform you're synthesizing or analyzing, you may find that other basis functions provide a better approximation with fewer terms. Wavelets are one example of alternative basis functions, and you can find links to additional information about wavelets on the book's website.

3.4 Wave packets and dispersion

Once you grasp the concepts of superposition and Fourier synthesis, you have the foundation in place to understand wave packets and dispersion. If you followed the developments in the previous section, you know that a periodic pulsetrain in space or time can be synthesized from the right mix of discrete frequency components, and even a single non-periodic pulse can be produced by a continuous function in the spatial or temporal frequency domain.

You may have noticed that the functions used to explain Fourier synthesis and analysis are typically distance-domain functions such as $X(x)$ or time-domain functions such as $T(t)$. But you know that propagating waves are defined by functions that involve both space and time, such as $f(x - vt)$. Does the full apparatus of Fourier series and Fourier analysis apply to such functions?

Happily, it does, but there are conditions which can lead to a somewhat more complicated picture when you apply Fourier concepts to propagating waves. Specifically, consider what might happen if the different frequency components that make up a resultant waveform don't propagate at the same speed. In the previous section you saw how to determine just the right amplitude and just the right frequency of sines and cosines to produce the desired resultant waveform. But consider the *phase* of those frequency components (odd-harmonic sine waves in the case of the square pulse). If you look back at the frequency components shown in Fig. 3.11, you'll see that each of them starts at the part of the cycle for which the value of the sine function is zero and then moves to positive values (called the positive-going zero crossing). So in that case, each of the frequency components had the same starting phase.

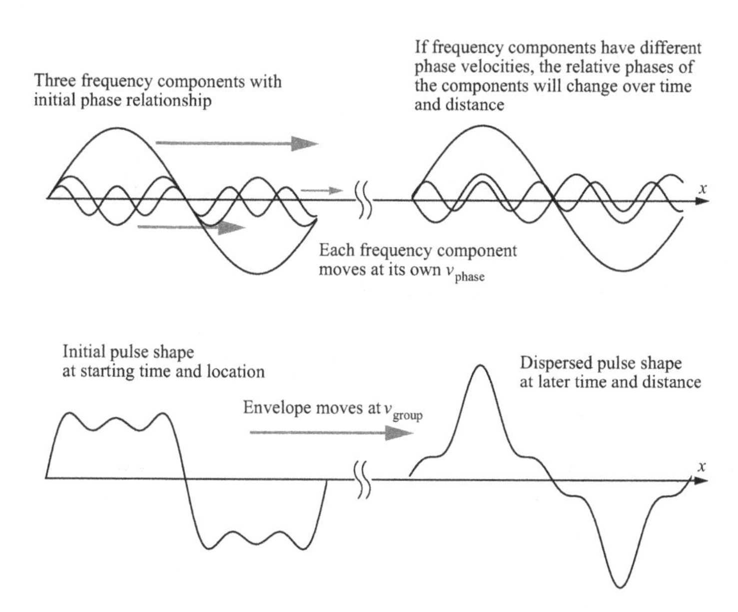

Figure 3.24 Differing frequency-component phase velocities produce dispersion.

Now imagine what happens as time goes by. If all of the frequency components travel at the same speed, the relative phase between the components remains the same, and the wave packet retains its shape (sometimes called the "envelope") and moves at the same speed as the individual frequency components.

But consider what would happen if the frequency components that make up the waveform (called a "wave packet" if it's localized in time and space) had different speeds. In that case, the relative phase between the components changes over distance, so they add up to give a different shape, and the speed of the packet's envelope will be different from the speed of the frequency components.

You can see an example of that in Fig. 3.24. In the left portion of this figure, three frequency components have the right amplitude and frequency to add up to a square(ish) pulse at some initial location and time. But as the three component waves propagate, if they move at different speeds, then at a different location the relative phase between the component waves will be different, which means they'll add up to a different resultant. You can see the result of that change in relative phase in the bottom-right portion of Fig. 3.24: The shape of the resultant waveform changes over distance.

This effect is called "dispersion". And when dispersion is present, the speed of each individual frequency component is called the phase velocity or phase speed of that component, and the speed of the wave packet's envelope is called the "group velocity" or group speed.[6]

Fortunately, there's a relatively straightforward way to determine the group velocity of a wave packet. To understand that, consider what happens when you add two wave components that have the same amplitude and slightly different frequency.

An example of this is shown in Fig. 3.25. As you can see in the bottom portion of this figure, the two component waves start off in phase, which means that they add constructively and produce a large resultant wave. But since the two component waves have slightly different frequencies, they soon get out of phase with respect to one another, and when their phase difference reaches 180° (that is, when they're completely out of phase), they cancel each other out and the amplitude of the resultant waveform becomes very small. But, as the two component waves continue to advance, at some point they once again have zero phase difference, and the resultant waveform gets large again.

When the amplitude of the resultant waveform varies as it does in Fig. 3.25, the waveform is said to be "modulated", and this particular type of modulation is called "beats". You can hear this effect by programming two sound generators to produce equally loud tones at slightly different frequencies; the volume of the resultant sound wave will vary between loud and soft at the "beat frequency" (which equals the frequency difference between the two component waveforms). Old-school piano tuners would simultaneously strike one of the piano's keys and a tuning fork and listen for the beats – the more closely the piano key's frequency matched that of the tuning fork, the slower the beats.

The modulation envelope shown in Fig. 3.25 provides a convenient way to determine the group velocity of a wave packet. To see how that works, write the phase of each of the two component waves as $\phi = kx - \omega t$, but remember that each wave has its own k and ω:

$$\phi_1 = k_1 x - \omega_1 t,$$
$$\phi_2 = k_2 x - \omega_2 t.$$

This means that the phase difference between the waves is

$$\Delta\phi = \phi_2 - \phi_1 = (k_2 x - \omega_2 t) - (k_1 x - \omega_1 t)$$
$$= (k_2 - k_1)x - (\omega_2 - \omega_1)t.$$

[6] Since velocity is a vector, phase and group velocity should include the direction, but in this context speed and velocity are used interchangeably.

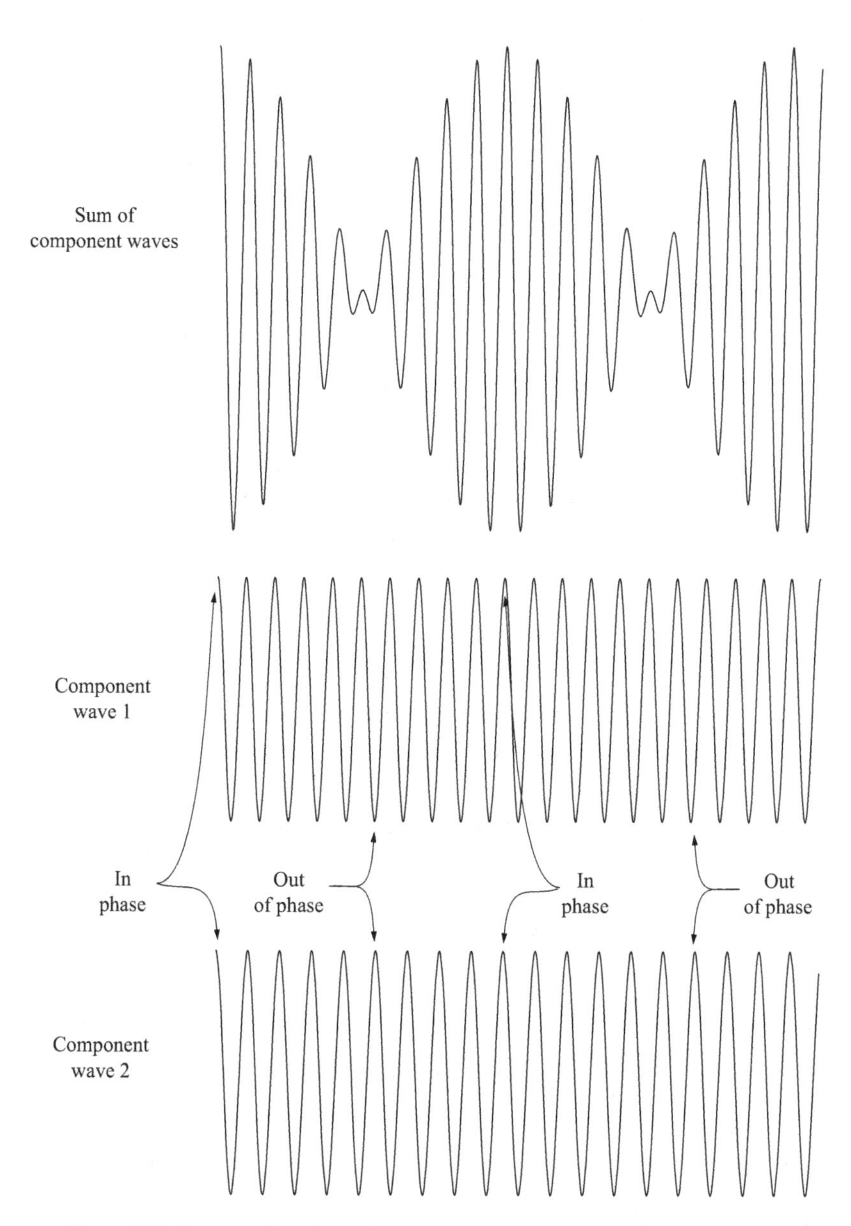

Figure 3.25 Beats produced by two component waves.

At a given value of the phase difference between the waves ($\Delta\phi$), the two waves add up to produce a certain value of the envelope of the resultant wave. To determine how fast that envelope moves, consider what happens over a small increment of time (Δt) and distance (Δx). Both component waves will move a certain distance in that time, but if you're following a point on the resultant wave, the relative phase between the two component waves must be the same. So whatever change occurs due to the passage of time Δt must be compensated for by a phase change due to a change in Δx. This means that

$$(k_2 - k_1)\Delta x = (\omega_2 - \omega_1)\Delta t$$

or

$$\frac{\Delta x}{\Delta t} = \frac{\omega_2 - \omega_1}{k_2 - k_1}. \tag{3.36}$$

But $\Delta x/\Delta t$ is the distance that the envelope moves divided by the time it takes, so this is the group velocity. And, although Eq. (3.36) is useful if you're dealing with only two waves, a far more general expression can be found for a group of waves with wavenumbers clustered around an average wavenumber k_a by expanding $\omega(k)$ in a Taylor series:

$$\omega(k) = \omega(k_a) + \left.\frac{d\omega}{dk}\right|_{k=k_a}(k - k_a) + \left.\frac{1}{2!}\frac{d^2\omega}{dk^2}\right|_{k=k_a}(k - k_a)^2 + \cdots .$$

For the case in which the difference between the wavenumbers is small, the higher-order terms of the expansion are negligible, which allows you to write

$$\omega(k) \approx \omega(k_a) + \left.\frac{d\omega}{dk}\right|_{k=k_a}(k - k_a)$$

or

$$\frac{\omega(k) - \omega(k_a)}{k - k_a} \approx \left.\frac{d\omega}{dk}\right|_{k=k_a}.$$

Thus

$$v_{\text{group}} = \frac{\omega(k) - \omega(k_a)}{k - k_a} \approx \left.\frac{d\omega}{dk}\right|_{k=k_a}.$$

So the group velocity of a wave packet is $v_{\text{group}} = d\omega/dk$ and the phase velocity of a wave component is $v_{\text{phase}} = \omega/k$.

When dealing with dispersion, you're very likely to encounter graphs in which ω is plotted on the vertical axis and k on the horizontal. If no dispersion is present, then the wave angular frequency ω is related to the wavenumber k by the equation $\omega = c_1 k$, where c_1 represents the speed of propagation, which is constant over all values of k. In this case, the dispersion plot is linear, as shown in Fig. 3.26.

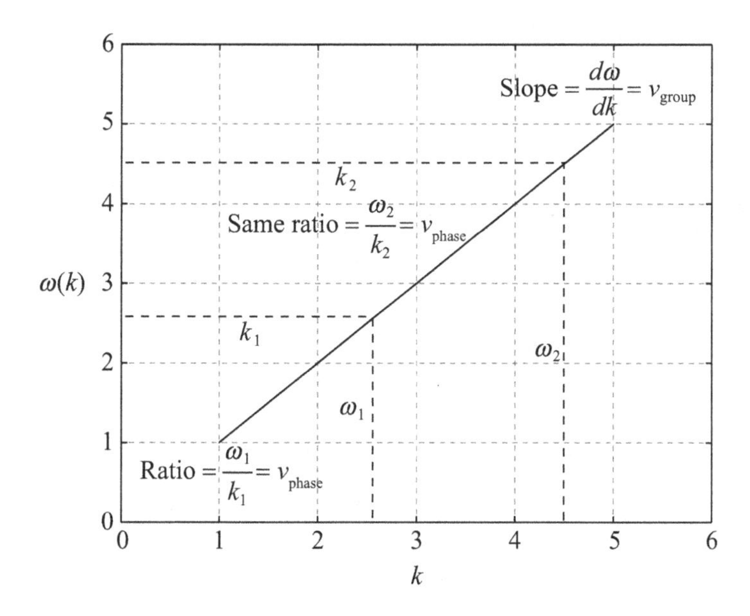

Figure 3.26 The linear dispersion relation for $\omega = c_1 k$.

In the non-dispersive case, the phase velocity ω/k is the same at all values of k and is the same as the group velocity $d\omega/dk$.

When dispersion is present, the relationship between the phase velocity of the component waves and the group velocity of the wave packet depends on the nature of the dispersion. In one important case pertaining to quantum waves (which you can read about in Chapter 6), the angular frequency is proportional to the square of the wavenumber ($\omega = c_2 k^2$).

You can see a plot of ω versus k for this case in Fig. 3.27. In this case, both the phase velocity and the group velocity increase as the wavenumber k increases. To find the phase velocity, use $v_{phase} = \omega/k$:

$$v_{phase} = \frac{\omega}{k} = \frac{c_2 k^2}{k} = c_2 k,$$

while for the group velocity

$$v_{group} = \frac{d\omega}{dk} = \frac{d(c_2 k^2)}{dk} = 2c_2 k,$$

which is twice the phase velocity. This is true at all values of k, as you can see by plotting v_{phase} and v_{group} vs. k, as in Fig. 3.28. Notice that both v_{phase} and v_{group} are increasing linearly with k, but v_{group} is always twice as big as v_{phase}.

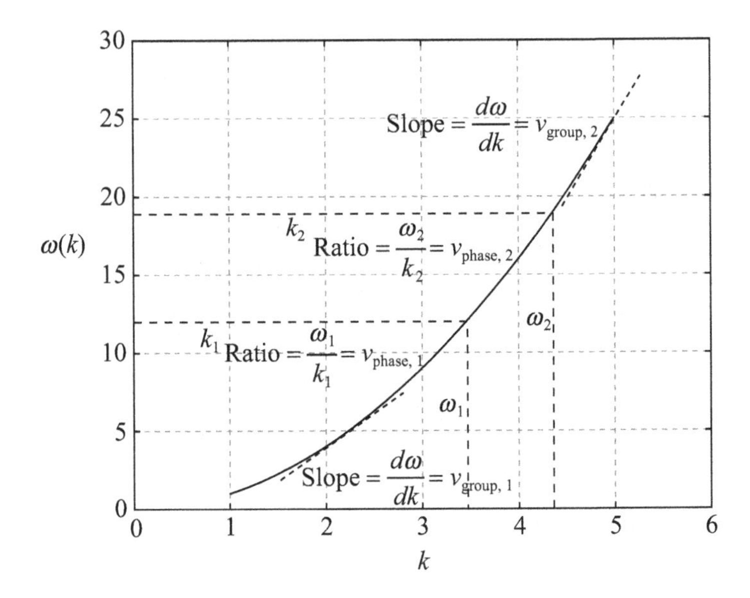

Figure 3.27 The dispersion relation for $\omega = c_2k^2$.

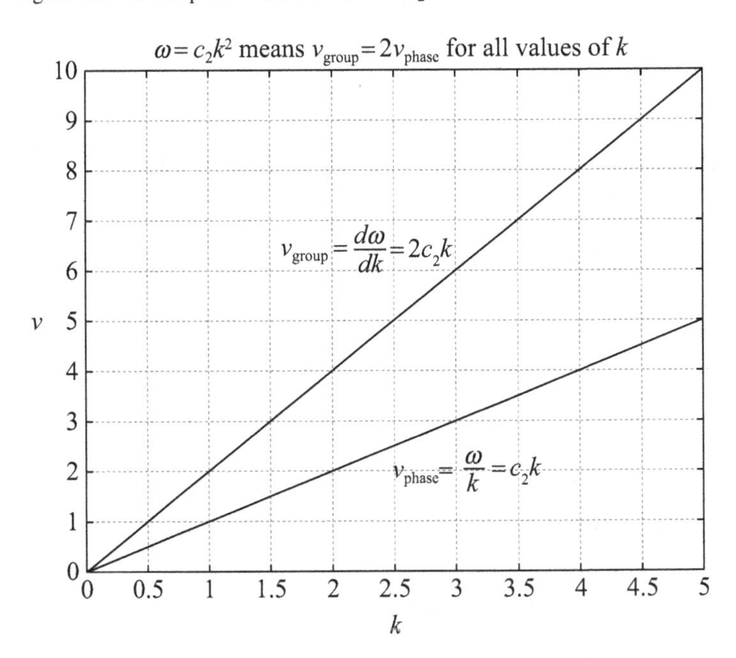

Figure 3.28 Phase and group velocity for $\omega = c_2k^2$.

3.5 Problems

3.1. Show that the expression $C\sin(\omega t + \phi_0)$ is equivalent to $A\cos(\omega t) + B\sin(\omega t)$, and write equations for C and ϕ_0 in terms of A and B.

3.2. Sketch the two-sided wavenumber spectrum of the function $X(x) = 6 + 3\cos(20\pi x - \pi/2) - \sin(5\pi x) + 2\cos(10\pi x + \pi)$.

3.3. Find the Fourier series representation of a periodic function for which one period is given by $f(x) = x^2$ for x between $-L$ and $+L$.

3.4. Verify the coefficients A_0, A_n, and B_n for the periodic triangle wave shown in Fig. 3.15.

3.5. If a string (fixed at both ends) is plucked rather than struck (non-zero initial displacement, zero initial velocity), show that the displacement at position x and time t is

$$y(x,t) = \sum_{n=1}^{\infty} B_n \sin\left(\frac{n\pi x}{L}\right)\cos\left(\frac{n\pi vt}{L}\right).$$

3.6. Find the B_n coefficients for the plucked string of the previous problem if the initial displacement is given by the function shown below:

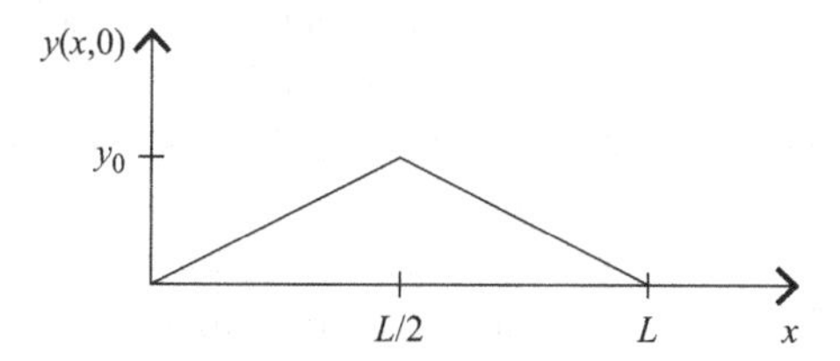

3.7. Find the B_n coefficients for a hammered string with initial displacement of zero and initial velocity given by the function shown below:

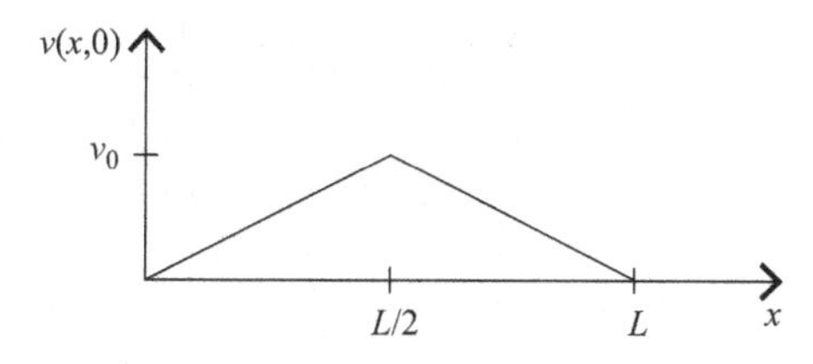

3.8. Find the Fourier transform of the Gaussian function $T(t) = \sqrt{\alpha/\pi}\, e^{-\alpha t^2}$.

3.9. Show that the complex-exponential version of the Fourier series (Eq. (3.31)) is equivalent to the version using sines and cosines (Eq. (3.25)).

3.10. Under certain conditions, the dispersion relation for deep-water waves is $\omega = \sqrt{gk}$, in which g is the acceleration due to gravity. Compare the group velocity with the phase velocity for this type of wave.

4

The mechanical wave equation

This chapter is the first of three that deal in detail with the wave equation for three different types of waves. Mechanical waves are described in this chapter, Chapter 5 discusses electromagnetic waves, and Chapter 6 is all about quantum-mechanical waves. You'll find frequent reference in these three chapters to concepts and equations contained in Chapters 1, 2, and 3, so if you've skipped those chapters but find yourself needing a bit more explanation of a particular idea, you can probably find it in one of the earlier chapters.

The layout of this chapter is straightforward. After an overview of the properties of mechanical waves in Section 4.1, you'll find a discussion of two types of mechanical waves: transverse waves on a string (Section 4.2) and longitudinal pressure waves (Section 4.3). Section 4.4 is about energy and power in mechanical waves, and Section 4.5 discusses the reflection and transmission of mechanical waves.

4.1 Properties of mechanical waves

We're surrounded by the electromagnetic waves of Chapter 5, and we're made of particles that can be described by the quantum-mechanical waves of Chapter 6, but when you ask most people to describe a wave, they think of the mechanical waves that are the subject of this chapter. That's probably because the "waving" of many types of mechanical waves can be readily observed. For mechanical waves, the waving is done by bits of mass – atoms, molecules, or connected sets of particles, which means that mechanical waves can exist only within a physical material (called the "medium" in which the wave propagates).

For mechanical waves such as a wave on a string, the disturbance of the wave is the physical displacement of matter from the undisturbed (equilibrium) position. In that case, the disturbance has dimensions of distance, and the SI units of the disturbance are meters. In other mechanical waves, such as pressure waves, you may encounter other measures of the wave's disturbance. For example, in sound waves, the disturbance may be measured as the change in the density of the medium or the change in the pressure within the medium from their equilibrium values. Thus you may see the "displacement" of pressure waves expressed as a density fluctuation (with SI units of kg/m^3) or as a pressure change (with SI units of N/m^2).

So one common aspect of all mechanical waves is that they require a physical medium to exist (the tag line for the 1979 science-fiction film *Alien* was correct: In space, no one can hear you scream). For any medium, two characteristics of the material have a critical impact on the propagation of mechanical waves. These characteristics are the material's "inertial" property and its "elastic" property.

The inertial property of the material is related to the mass (for discrete media) or mass density (for continuous media) of the material. Remember that "inertia" describes the tendency of all mass to resist acceleration. So material with high mass is more difficult to get moving (and more difficult to slow down once it's moving) than material with low mass. As you'll see later in this chapter, the mass density of the medium affects the velocity of propagation (and the dispersion) of mechanical waves as well as the ability of a source to couple energy into a wave within the medium (called the "impedance" of the medium).

The elastic property of the material is related to the restoring force that tends to return displaced particles to their equilibrium positions. So a medium can be elastic in the same way a rubber band or a spring is elastic – when you stretch it, displacing particles from their equilibrium positions, the medium produces restoring forces (if it didn't, you couldn't establish a mechanical wave within the medium). It may help you to think of this as a measure of the "stiffness" of the medium: Stiff rubber bands and springs are harder to stretch because the restoring forces are stronger. As you might expect, the strength of those restoring forces joins the mass density in determining the speed of propagation, the dispersion, and the impedance of a medium.

Before considering the wave equation for mechanical waves, you should understand the difference between the motion of individual particles and the motion of the wave itself. Although the medium is disturbed as a wave goes by, which means that the particles of the medium are displaced from their equilibrium positions, those particles don't travel very far from their undisturbed positions. The particles oscillate about their equilibrium positions, but the wave does not carry the particles along – a wave is not like a steady breeze or an ocean current which transports material in bulk from one location to another. For mechanical waves, the net displacement of material produced by the wave over one cycle, or over one million cycles, is zero. So, if the particles aren't being carried along with the wave, what actually moves at the speed of the wave? As you'll see in Section 4.4, the answer is energy.

Although the displacement of individual particles within the medium is small, the direction of the particles' movement is important, because the direction of particle motion is often used to categorize the wave as either transverse or longitudinal (although in some waves, including ocean waves, the particles have both transverse and longitudinal motion). In a transverse wave, the particles move in a direction that is perpendicular to the direction of wave motion. That means that, for a transverse wave traveling to the right on this page, each particle would move up and down or in and out of the page, but not left or right. In a longitudinal wave, the particles of the medium move in a direction parallel and antiparallel to the direction of wave motion. So, for a longitudinal wave traveling to the right on this page, the particles would move to the right and left. Which type of wave is set up in a given medium depends on the source and on the direction of the restoring force, as you can see in the discussion of transverse string waves in Section 4.2 and longitudinal pressure waves in Section 4.3.

4.2 Waves on a string

If you've done any reading about mechanical waves in other books or online, it's very likely that you encountered a discussion of transverse waves on a stretched string (or spring). In the most common scenario, a horizontal string is struck or plucked somewhere along its length, causing a vertical displacement of some portion of the string from the equilibrium position. You may also have seen a case in which the wave motion is induced by driving one end of the string up and down. The following analysis applies to any of these situations.

There are several different ways to analyze this type of motion, but we think the most straightforward is to use Newton's second law to relate the tension forces on a segment of the string to the acceleration of that segment. Through that analysis, you can arrive at a version of the classical wave equation that shows how the segment will move, and how the inertial and elastic properties of the string determine the speed of the induced waves.

To see how that works, consider a segment of a string with uniform linear density (mass per unit length) which has been displaced from the equilibrium (horizontal) position, as shown in Fig. 4.1. The string has elasticity, so, as the

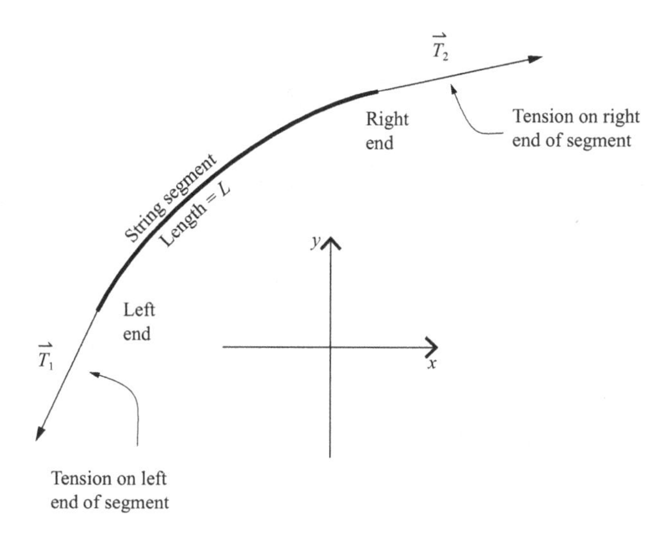

Figure 4.1 Tension forces on a segment of string.

segment is displaced, there are tension forces acting on each end, which tend to return the segment to its equilibrium position. So the string is really acting like a spring that has been stretched.

As the string stretches, the mass per unit length must decrease, but we're going to consider the case in which that change in linear density due to that stretching is negligible. We'll also consider vertical displacements that are small relative to the horizontal extent of the disturbance, so the angle (θ) that any portion of the segment makes with the horizontal direction is small.[1] And we'll assume that the effects of all other forces (such as gravity) are negligible compared with those of the tension forces.

As indicated in Fig. 4.1, the string's undisturbed (horizontal) position is along the x-axis and the displacement is in the y-direction. As the segment of string is displaced from its equilibrium position, the tension force $\vec{T}_1$ acts on the left end and the tension force $\vec{T}_2$ acts on the right end. You can't say much about the strength of these forces without knowing the elasticity of the string and the amount of displacement, but you can find some interesting physics even without that information. To see that, consider the x- and y-components of the tension forces $\vec{T}_1$ and $\vec{T}_2$ shown in Fig. 4.2.

In Fig. 4.2(a) the angle of the left end of the string segment from the horizontal is shown as θ_1, and the x- and y-components of tension force $\vec{T}_1$

[1] How small? Small enough to allow us to make approximations such as $\cos\theta \approx 1$ and $\sin\theta \approx \tan\theta$, which are good to within 10% if θ is less than 25°.

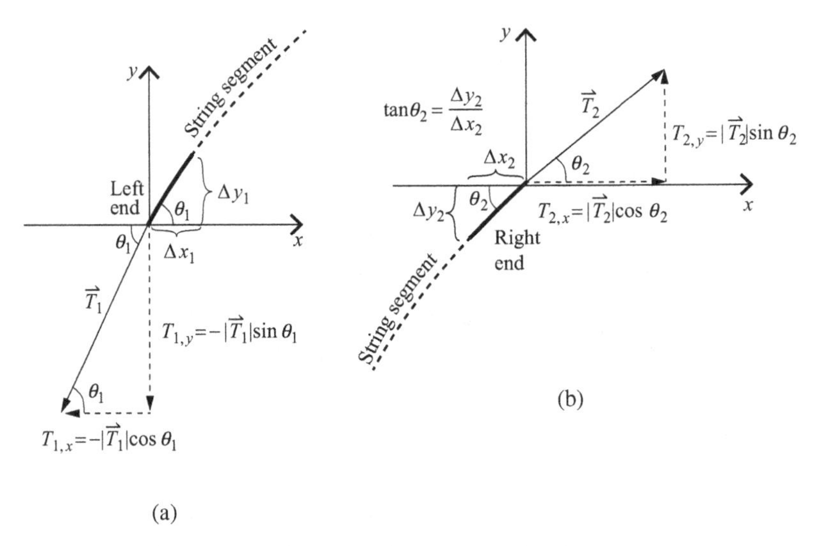

Figure 4.2 Components of left-end and right-end tension forces.

can be seen to be

$$T_{1,x} = -|\vec{T}_1|\cos\theta_1,$$
$$T_{1,y} = -|\vec{T}_1|\sin\theta_1.$$

Likewise, in Fig. 4.2(b) the angle of the right end of the string segment from the horizontal is shown as θ_2, and the x- and y-components of tension force $\vec{T}_2$ can be seen to be

$$T_{2,x} = |\vec{T}_2|\cos\theta_2,$$
$$T_{2,y} = |\vec{T}_2|\sin\theta_2.$$

The next step is to use Newton's second law to write the sum of the forces acting on the segment in the x-direction as the mass (m) of the segment times the segment's x-acceleration (a_x),

$$\Sigma F_x = -|\vec{T}_1|\cos\theta_1 + |\vec{T}_2|\cos\theta_2 = ma_x, \tag{4.1}$$

and likewise for the y-components of force and acceleration,

$$\Sigma F_y = -|\vec{T}_1|\sin\theta_1 + |\vec{T}_2|\sin\theta_2 = ma_y. \tag{4.2}$$

As long as the string segment oscillates up and down along the y-direction, you can take $a_x = 0$, and, if the amplitude of the oscillation is small, then θ_1

and θ_2 are both small, and

$$\cos\theta_1 \approx \cos\theta_2 \approx 1.$$

Using these approximations and setting a_x to zero gives

$$\Sigma F_x \approx -|\vec{T}_1|(1) + |\vec{T}_2|(1) = 0,$$

which means that

$$|\vec{T}_1| \approx |\vec{T}_2|, \tag{4.3}$$

so the magnitudes of the tension forces on each end of the string are approximately equal. But the directions of those forces (and thus their y-components) are not equal, and the difference in those components has important consequences for the motion of the segment.

To see that, take a look at the portions of Fig. 4.2 that show the slopes of the segment at its left and right ends. As shown in the figure, those slopes are given by the equations

$$\begin{aligned} \text{Left-end slope} &= \tan\theta_1 = \frac{\Delta y_1}{\Delta x_1}, \\ \text{Right-end slope} &= \tan\theta_2 = \frac{\Delta y_2}{\Delta x_2}. \end{aligned} \tag{4.4}$$

And here's the trick that moves this analysis in the direction of the classical wave equation. At the left end of the segment, consider only an infinitesimal piece of the string by letting Δx_1 approach zero. When you do that, the slope approaches the partial derivative of y with respect to x evaluated at the left end of the segment:

$$\frac{\Delta y_1}{\Delta x_1} \rightarrow \left[\frac{\partial y}{\partial x}\right]_{\text{left}}.$$

Now do the same thing at the right end by letting Δx_2 approach zero:

$$\frac{\Delta y_2}{\Delta x_2} \rightarrow \left[\frac{\partial y}{\partial x}\right]_{\text{right}}.$$

But Eqs. (4.4) say that the slopes at the ends of the segment are equal to the tangent of θ_1 and θ_2, and for small angles those tangents may be approximated by sines:

$$\tan\theta_1 = \frac{\sin\theta_1}{\cos\theta_1} \approx \sin\theta_1,$$
$$\tan\theta_2 = \frac{\sin\theta_2}{\cos\theta_2} \approx \sin\theta_2.$$

This means that the sine terms in Eq. (4.2) may be approximated by tangents, and the tangents are equivalent to partial derivatives. Thus the sum of the forces in the y-direction may be written as

$$\Sigma F_y \approx -|\vec{T}_1| \left[\frac{\partial y}{\partial x}\right]_{\text{left}} + |\vec{T}_2| \left[\frac{\partial y}{\partial x}\right]_{\text{right}} = ma_y$$

and, since $a_y = \partial^2 y/\partial t^2$, this is

$$m\frac{\partial^2 y}{\partial t^2} = -|\vec{T}_1| \left[\frac{\partial y}{\partial x}\right]_{\text{left}} + |\vec{T}_2| \left[\frac{\partial y}{\partial x}\right]_{\text{right}}.$$

But in the small-amplitude and uniform-density approximations, Eq. (4.3) says that the magnitudes of the tensions on the left and right ends of the segment are equal, so you can write $|\vec{T}_1| = |\vec{T}_2| = T$:

$$m\frac{\partial^2 y}{\partial t^2} = T \left[\frac{\partial y}{\partial x}\right]_{\text{right}} - T \left[\frac{\partial y}{\partial x}\right]_{\text{left}}$$

or

$$\frac{\partial^2 y}{\partial t^2} = \frac{T}{m} \left\{ \left[\frac{\partial y}{\partial x}\right]_{\text{right}} - \left[\frac{\partial y}{\partial x}\right]_{\text{left}} \right\}.$$

If you look at the term inside the large curly brackets, you'll see that it's just the change in the slope of the segment between the left end and the right end. That change can be written using Δ notation as

$$\left\{ \left[\frac{\partial y}{\partial x}\right]_{\text{right}} - \left[\frac{\partial y}{\partial x}\right]_{\text{left}} \right\} = \Delta\left(\frac{\partial y}{\partial x}\right),$$

which means

$$\frac{\partial^2 y}{\partial t^2} = \frac{T}{m} \Delta\left(\frac{\partial y}{\partial x}\right).$$

The final steps to the wave equation can be made by considering the mass (m) of the string segment. If the linear density of the string is μ and the length of the segment is L, the mass of the segment is μL. But as long as the amplitude of the displacement is small, then $L \approx \Delta x$, so the mass of the segment can be written as $m = \mu\,\Delta x$. Thus

$$\frac{\partial^2 y}{\partial t^2} = \frac{T}{\mu \, \Delta x} \Delta \left(\frac{\partial y}{\partial x}\right).$$

But, for small Δx,

$$\frac{\Delta(\partial y/\partial x)}{\Delta x} = \frac{\partial^2 y}{\partial x^2},$$

which means

$$\frac{\partial^2 y}{\partial t^2} = \frac{T}{\mu} \left(\frac{\partial^2 y}{\partial x^2}\right)$$

or

$$\frac{\partial^2 y}{\partial x^2} = \frac{\mu}{T} \left(\frac{\partial^2 y}{\partial t^2}\right). \tag{4.5}$$

If you were tracking the developments of Chapter 2, this should look very familiar. A second-order spatial derivative on the left and a second-order time derivative on the right (with no other terms except for a constant multiplicative factor on the right) mean that this equation has the same form as the classical wave equation (Eq. (2.5)), and all the analysis discussed in Chapters 2 and 3 can be applied to this waving string.

One interesting observation about this form of the wave equation is that, since the displacement (y) in this case is actually a physical displacement,[2] the term on the left side of Eq. (4.5) ($\partial^2 y/\partial x^2$) represents the curvature of the string. Hence the classical wave equation, when applied to a string, tells you that the acceleration of any segment of the string ($\partial^2 y/\partial t^2$) is proportional to the curvature of that segment. This should make sense in light of the relationship between the slopes at the left and right ends of the segment and the y-components of the tension. With no curvature, the slopes at the two ends would be equal, and the y-components of the tension would be equal and opposite. Thus the acceleration of the segment would be zero.

As you may have surmised, you can determine the phase speed of the wave by comparing the multiplicative term in Eq. (2.5) with that in Eq. (4.5). Setting these factors equal to one another gives

$$\frac{1}{v^2} = \frac{\mu}{T},$$

so

$$v = \sqrt{\frac{T}{\mu}}. \tag{4.6}$$

[2] Recall from Chapter 1 that "displacement" may refer to any deviation from equilibrium, not just physical distance. But in this case, y is the actual distance from the equilibrium position.

As expected, the phase speed of the string wave depends both on the elastic (T) and on the inertial (μ) properties of the string (which is the medium of propagation in this case). Specifically, the higher the tension in the string, the faster the components of the wave will propagate (since T is in the numerator), and the higher the density of the string, the slower those components will move (since μ is in the denominator). So if string A has twice the tension and twice the linear mass density of string B, the phase velocity of the waves in the strings will be the same.

It's important to recognize that the velocity given by Eq. (4.6) is the speed of the wave along the string (that is, in the horizontal direction), not the speed of the segment in the vertical direction. That (transverse) speed is given by $\partial y/\partial t$, which you can determine by taking the time derivative of the wavefunction $y(x, t)$ that solves the wave equation.

And how do you find the wavefunction $y(x, t)$ that solves the wave equation? By applying the appropriate initial and/or boundary conditions for the problem you're trying to solve. You can see some examples of that in Sections 3.1 and 3.2 of Chapter 3. In that discussion, the phase speed of the waves was simply called "v", but now you know that speed to be $v = \sqrt{T/\mu}$.

The harmonic-function (sine and cosine) solutions to the classical wave equation can be very instructive in the case of transverse waves on a string, as you can see in the following example.

Example 4.1 *Compare the displacement, velocity, and acceleration for a transverse harmonic wave on a string.*

If the displacement $y(x, t)$ is given by $A\sin(kx - \omega t)$, the transverse velocity of any segment of the string is given by $v_t = \partial y/\partial t = -A\omega\cos(kx - \omega t)$, and the transverse acceleration is $a_t = \partial^2 y/\partial t^2 = -A\omega^2\sin(kx - \omega t)$. Note that this wave is moving in the positive x-direction, since the sign of the kx term is opposite to the sign of the ωt term. Plotting the displacement, transverse velocity, and transverse acceleration on the same graph, as in Fig. 4.3, reveals some interesting aspects of wave behavior. This figure is a snapshot of y, v_t, and a_t at time $t = 0$, and the angular frequency has been taken as $\omega = 1$ in order to scale all three waveforms to the same vertical size.

The reason for putting all three waveforms on the same graph is to help you see the relationship between the displacement, speed, and acceleration at any location (and at the same time). Consider, for example, the segment of the string at location x_1, for which the displacement has its maximum positive value (the first positive peak of the $y(x, t)$ waveform). At the instant ($t = 0$) when that segment is reaching maximum displacement, the velocity plot shows

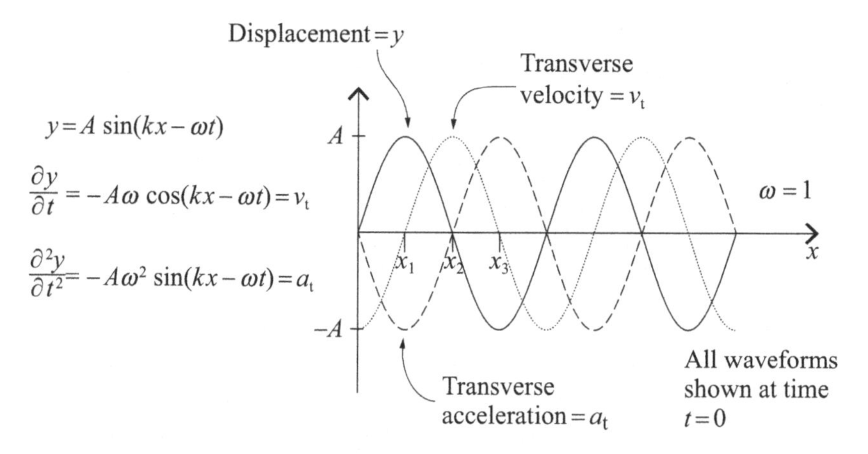

Figure 4.3 Time derivatives of a harmonic wave.

that the transverse velocity of that segment is zero. That's because the segment has instantaneously stopped at the top of its excursion as it transitions from moving upward (away from the equilibrium position) to moving downward (back toward the equilibrium position).

Now look at the acceleration plot for that same segment. At the same instant as the displacement has its maximum positive value and the transverse velocity is zero, the transverse acceleration of the segment has its maximum negative value. That makes sense, because, when the segment is at maximum displacement from equilibrium, the restoring force (the tension) is at its strongest, and the direction of that force is back toward equilibrium (downward when the segment is above the equilibrium position). Since the force on the segment is proportional to the segment's acceleration, maximum negative force means maximum negative acceleration.

Next, consider the segment of the string at position x_2. At time $t = 0$, that segment is passing through the equilibrium position, which you know by the fact that its displacement (y) is zero. Looking at that segment's velocity graph, you can see that it's reaching a peak – the velocity of that segment has its maximum positive value as the segment passes through equilibrium. As it does so, the tension force is entirely horizontal (it has no y-component), so the transverse acceleration (a_y) of the segment is zero.

The same analysis can be applied to the segment at position x_3, which has maximum negative displacement, zero transverse velocity, and maximum positive acceleration.

One mistake some students make when analyzing a graph such as Fig. 4.3 is to conclude that the segment of the string at position x_2 is moving from positive (above equilibrium) displacement to negative (below equilibrium) displacement. But remember that this wave is moving to the right, and you're looking at a snapshot of the wave taken at time $t = 0$. So, at a later instant in time, the wave will have moved to the right, and the segment at position x_2 will be *above* the equilibrium position – it will have positive displacement ($y > 0$). That's why the velocity of the segment at x_2 has its maximum positive value at the instant shown in the graph; the segment is moving in the positive y-direction.

Many students find that a good understanding of transverse waves on a string serves as a foundation for understanding a variety of other types of wave motion, including longitudinal pressure waves (described in Section 4.3) and even quantum waves (which are the subject of Chapter 6). One additional concept that you may find helpful is the effect of changes in the linear mass density (μ) or tension (T) of the string – in other words, how would the analysis change if the string were inhomogeneous?

You can see the effect of allowing the string characteristics to change with distance by treating both the density and the tension as functions of x: $\mu = \mu(x)$ and $T = T(x)$. In that case, a similar analysis to that shown above leads to this modified version of the wave equation:

$$\frac{\partial^2 y}{\partial t^2} = \frac{1}{\mu(x)} \frac{\partial}{\partial x} \left[T(x) \frac{\partial y}{\partial x} \right].$$

Notice that the tension $T(x)$ can not be moved out of the spatial derivative, and the ratio of the tension to the density may no longer be constant. In general, the spatial portions of the solutions to this equation are non-sinusoidal, which is why you may have read that sinusoidal spatial wavefunctions are the hallmark of a homogeneous medium. This has an interesting analogy with quantum waves in an inhomogeneous medium.

4.3 Pressure waves

Many of the important concepts and techniques used to analyze transverse waves on a string can also be applied to longitudinal pressure waves. Although there doesn't seem to be standard terminology for this kind of wave, by "pressure wave" we mean any wave in a medium in which a mechanical source (such as the vibrating tyne of a tuning fork, an oscillating piston, or the moving diaphragm of a speaker) causes a physical displacement and compression or rarefaction of the material in the direction in which the wave is propagating.

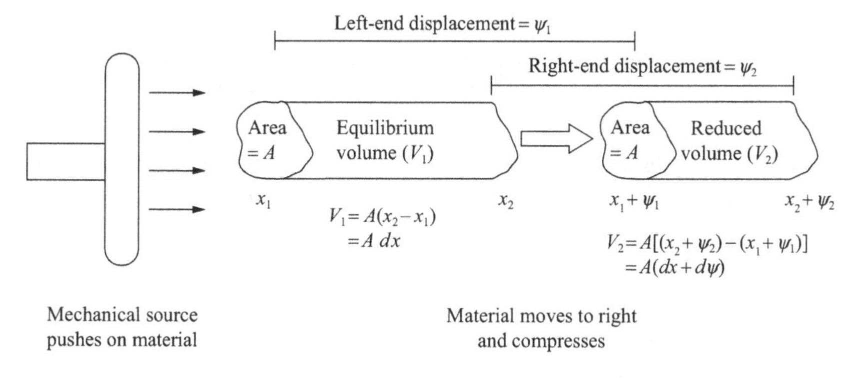

Figure 4.4 Displacement and compression of a segment of material.

You can see an illustration of how a pressure wave works in Fig. 4.4. As the mechanical wave source moves through the medium, it pushes on a nearby segment of the material, and that segment moves away from the source and is compressed (that is, the same amount of mass is squeezed into a smaller volume, so the density of the segment increases). That segment of increased density exerts pressure on adjacent segments, and in this way a pulse (if the source gives a single push) or a harmonic wave (if the source oscillates back and forth) is generated by the source and propagates through the material.

The "disturbance" of such waves involves three things: the longitudinal displacement of material, changes in the density of the material, and variation of the pressure within the material. So pressure waves could also be called "density waves" or even "longitudinal displacement waves", and when you see graphs of the wave disturbance in physics and engineering textbooks, you should make sure you understand which of these quantities is being plotted as the "displacement" of the wave.

As you can see in Fig. 4.4, we're still considering one-dimensional wave motion (that is, the wave propagates only along the x-axis). But pressure waves exist in a three-dimensional medium, so instead of considering the linear mass density μ (as we did for the string in the previous section), in this case it's the volumetric mass density ρ that will provide the inertial characteristic of the medium. But just as we restricted the string motion to small angles and considered only the transverse component of the displacement, in this case we'll assume that the pressure and density variations are small relative to the equilibrium values and consider only longitudinal displacement (so the material is compressed or rarefied only by changes in the segment length in the x-direction).

The most straightforward route to finding the wave equation for this type of wave is very similar to the approach used for transverse waves on a string, which means you can use Newton's second law to relate the acceleration of a segment of the material to the sum of the forces acting on that segment. To do that, start by defining the pressure (P) at any location in terms of the equilibrium pressure (P_0) and the incremental change in pressure produced by the wave (dP):

$$P = P_0 + dP.$$

Likewise, the density (ρ) at any location can be written in terms of the equilibrium density (ρ_0) and the incremental change in density produced by the wave ($d\rho$):

$$\rho = \rho_0 + d\rho.$$

Before relating these quantities to the acceleration of material in the medium using Newton's second law, it's worthwhile to familiarize yourself with the terminology and equations of volume compressibility. As you might imagine, when external pressure is applied to a segment of material, how much the volume (and thus the density) of that material changes depends on the nature of the material. To compress a volume of air by one percent requires a pressure increase of about one thousand Pa (pascals, or N/m^2), but to compress a volume of steel by one percent requires a pressure increase of more than one billion Pa. The compressibility of a substance is the inverse of its "bulk modulus" (usually written as K or B, with units of pascals), which relates an incremental change in pressure (dP) to the fractional change in density ($d\rho/\rho_0$) of the material:

$$K \equiv \frac{dP}{d\rho/\rho_0} \tag{4.7}$$

or

$$dP = K\frac{d\rho}{\rho_0}. \tag{4.8}$$

With this relationship in hand, you're ready to consider Newton's second law for the segment of material being displaced and compressed (or rarefied) by the wave. To do that, consider the pressure from the surrounding material acting on the left and on the right side of the segment, as shown in Fig. 4.5.

Notice that the pressure (P_1) on the left end of the segment is pushing in the positive x-direction and the pressure on the left end of the segment is pushing in the negative x-direction. Setting the sum of the x-direction forces equal to the acceleration in the x-direction gives

$$\sum F_x = P_1A - P_2A = ma_x, \tag{4.9}$$

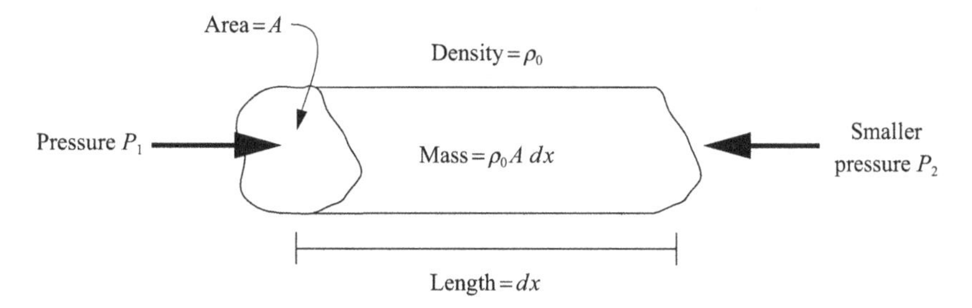

Figure 4.5 Pressure on a segment of material.

where m is the mass of the segment. If the cross-sectional area of the segment is A and the length of the segment is dx, the volume of the segment is $A\,dx$, and the mass of the segment is this volume times the equilibrium density of the material:

$$m = \rho_0 A\,dx.$$

Notice also that the pressure on the right end of the segment is smaller than the pressure on the left end, since the source is pushing on the left end, which means that the acceleration at this instant will be toward the right. Using the symbol ψ to represent the displacement of the material due to the wave, the acceleration in the x-direction can be written as

$$a_x = \frac{\partial^2 \psi}{\partial t^2}.$$

Substituting these expressions for m and a_x into Newton's second law (Eq. (4.9)) gives

$$\sum F_x = P_1 A - P_2 A = \rho_0 A\,dx \frac{\partial^2 \psi}{\partial t^2}.$$

Writing the pressure P_1 at the left end as $P_0 + dP_1$ and the pressure P_2 at the right end as $P_0 + dP_2$ means that

$$\begin{aligned} P_1 A - P_2 A &= (P_0 + dP_1)A - (P_0 + dP_2)A \\ &= (dP_1 - dP_2)A. \end{aligned}$$

But the change in dP (that is the change in the overpressure (or underpressure) produced by the wave) over the distance dx can be written as

$$\text{Change in overpressure} = dP_2 - dP_1 = \frac{\partial(dP)}{\partial x}\,dx,$$

which means

$$-\frac{\partial(dP)}{\partial x}\,dx\,A = \rho_0 A\,dx\,\frac{\partial^2\psi}{\partial t^2}$$

or

$$\rho_0\frac{\partial^2\psi}{\partial t^2} = -\frac{\partial(dP)}{\partial x}.$$

But $dP = d\rho K/\rho_0$, so

$$\rho_0\frac{\partial^2\psi}{\partial t^2} = -\frac{\partial[(K/\rho_0)d\rho]}{\partial x}. \tag{4.10}$$

The next step is to relate the change in density ($d\rho$) to the displacements of the left and right ends of the segment (ψ_1 and ψ_2). To do that, note that the mass of the segment is the same before and after the segment is compressed. That mass is the segment's density times its volume ($m = \rho V$), and the volume of the segment can be seen in Fig. 4.4 to be $V_1 = A\,dx$ before compression and $V_2 = A(dx + d\psi)$ after compression. Thus

$$\rho_0 V_1 = (\rho_0 + d\rho)V_2,$$
$$\rho_0(A\,dx) = (\rho_0 + d\rho)A(dx + d\psi).$$

The change in displacement ($d\psi$) over distance dx can be written as

$$d\psi = \frac{\partial\psi}{\partial x}\,dx,$$

so

$$\rho_0(A\,dx) = (\rho_0 + d\rho)A\left(dx + \frac{\partial\psi}{\partial x}\,dx\right),$$
$$\rho_0 = (\rho_0 + d\rho)\left(1 + \frac{\partial\psi}{\partial x}\right)$$
$$= \rho_0 + d\rho + \rho_0\frac{\partial\psi}{\partial x} + d\rho\,\frac{\partial\psi}{\partial x}.$$

Since we're restricting our consideration to cases in which the density change ($d\rho$) produced by the wave is small relative to the equilibrium density (ρ_0), the term $d\rho\,\partial\psi/\partial x$ must be small compared with the term $\rho_0\,\partial\psi/\partial x$. Thus to a reasonable approximation we can write

$$d\rho = -\rho_0\frac{\partial\psi}{\partial x},$$

which we can insert into Eq. (4.10), giving

$$\rho_0 \frac{\partial^2 \psi}{\partial t^2} = -\frac{\partial[(K/\rho_0)(-\rho_0\, \partial\psi/\partial x)]}{\partial x}$$
$$= \frac{\partial[K(\partial\psi/\partial x)]}{\partial x}.$$

Rearranging makes this into an equation with a familiar form:

$$\rho_0 \frac{\partial^2 \psi}{\partial t^2} = K \frac{\partial^2 \psi}{\partial x^2}$$

or

$$\frac{\partial^2 \psi}{\partial x^2} = \frac{\rho_0}{K} \frac{\partial^2 \psi}{\partial t^2}. \tag{4.11}$$

As in the case of transverse waves on a string, you can determine the phase speed of a pressure wave by comparing the multiplicative term in the classical wave equation (Eq. (2.5)) with that in Eq. (4.11). Setting these factors equal to one another gives

$$\frac{1}{v^2} = \frac{\rho_0}{K}$$

so

$$v = \sqrt{\frac{K}{\rho_0}}. \tag{4.12}$$

As expected, the phase speed of the pressure wave depends both on the elastic (K) and on the inertial (ρ_0) properties of the medium. Specifically, the higher the bulk modulus of the material (that is, the stiffer the material), the faster the components of the wave will propagate (since K is in the numerator), and the higher the density of the medium, the slower those components will move (since ρ_0 is in the denominator).

Example 4.2 *Determine the speed of sound in air.*

Sound is a type of pressure wave, so you can use Eq. (4.12) to determine the speed of sound in air, if you know the values of the bulk modulus and density of air. Since the air pressure may be more readily available than the bulk modulus in the region of interest, you may find it helpful to write this equation in a form that explicitly includes pressure.

To do that, use the definition of the bulk modulus (Eq. (4.7)) to write Eq. (4.12) as

$$v = \sqrt{\frac{K}{\rho_0}} = \sqrt{\frac{dP/(d\rho/\rho_0)}{\rho_0}} = \sqrt{\frac{dP}{d\rho}}. \tag{4.13}$$

The quantity $dP/d\rho$ can be related to the equilibrium pressure (P_0) and density (ρ_0) using the adiabatic gas law. Since an adiabatic process is one in which energy does not flow between a system and its environment by heat, using the adiabatic law means that we're assuming that the regions of compression and rarefaction produced by the sound wave will not lose or gain energy by heating as the wave oscillates. That's a good assumption for sound waves in air under typical conditions, because the flow of energy by conduction (molecules colliding and transferring kinetic energy) occurs over distances comparable to the mean free path (the average distance molecules travel between collisions). That distance is several orders of magnitude smaller than the distance between regions of compression and rarefaction (that is, half a wavelength) in sound waves. So the squeezing and stretching of the air by the wave produces regions of slightly higher and slightly lower temperature, and the molecules do not move far enough to restore equilibrium before the wave causes the compressed regions to rarefy and the rarefied regions to compress. Thus the wave action may indeed be considered to be an adiabatic process.[3]

To apply the adiabatic gas law, write the relationship between pressure (P) and volume (V) as

$$PV^{\gamma} = \text{constant}, \tag{4.14}$$

in which γ represents the ratio of specific heats at constant pressure and constant volume and has a value of approximately 1.4 for air under typical conditions.

Since volume is inversely proportional to density ρ, Eq. (4.14) can be written as

$$P = (\text{constant})\rho^{\gamma},$$

so

$$\frac{dP}{d\rho} = (\text{constant})\gamma\rho^{\gamma-1} = \gamma\frac{(\text{constant})\rho^{\gamma}}{\rho},$$

but $(\text{constant})\rho^{\gamma} = P$, so this is

$$\frac{dP}{d\rho} = \gamma\frac{P}{\rho}.$$

Inserting this into Eq. (4.13) gives

[3] When Newton first calculated the speed of sound in his great *Principia*, he instead used a constant-temperature law (Boyle's law), which caused him to underestimate the speed of sound by about 15%.

$$v = \sqrt{\gamma \frac{P}{\rho}}.$$

For typical values for air of $P = 1 \times 10^5$ Pa and $\rho = 1.2$ kg/m^3, this yields a value for the speed of sound of

$$v = \sqrt{1.4 \frac{1 \times 10^5}{1.2}} = 342 \text{ m/s},$$

which is very close to the measured value.

4.4 Energy and power of mechanical waves

If you've read Section 4.1, you may have noticed the answer to the question of exactly what's propagating in a mechanical wave (it's not the particles of the medium, since the particles are not carried along with the wave). That answer is "energy". In this section, you'll learn how the amount of energy and the rate of energy flow in a mechanical wave are related to the parameters of the wave.

As you may recall from introductory physics, the mechanical energy of a system consists of kinetic energy (often characterized as "energy of motion") and potential energy (often characterized as "energy of position"). You may also recall that the kinetic energy of a moving object is proportional to the mass of the object and the square of the object's speed, while the object's potential energy depends on its position in a region in which external forces act upon the object.

Potential energy comes in several types, and which type is relevant for a given problem depends on the nature of the forces acting on the object. An object in a gravitational field (such as that produced by a star or planet) has gravitational potential energy, and an object being acted upon by an elastic force (such as that produced by a spring or stretched string) has elastic potential energy. Another bit of introductory physics you should keep in mind concerns conservative forces. As you may recall, a conservative force is a force for which the work done by the force as the object changes position depends only on the change in position and not on the path taken by the object. Gravity and elastic forces are conservative, while forces such as friction and drag are non-conservative, because, for such dissipative forces, longer paths convert more mechanical energy into internal energy, and you don't get that energy back by going over the same path in the opposite direction. When a conservative force acts on an object, the change in the object's potential energy as the object changes position equals the work done by that force. Finally, recall that you're free to choose the reference value (the location at which the potential energy equals zero) wherever you like, so the value of the potential energy is arbitrary – it's only the *change* in the potential energy that has any physical significance.

Applying the concepts of kinetic and potential energy to a mechanical wave such as a transverse wave on a string is straightforward. The most common approach is to find expressions for the kinetic and potential energies of one small segment of the string, which give the energy density (that is, the energy per unit length) of the string. For harmonic waves, you can then find the total energy in each wavelength by adding the kinetic and potential energy densities and integrating the result over a distance equal to one wavelength.

As mentioned above, the kinetic energy (KE) of a small segment of the string depends on the segment's mass (m) and the square of the segment's transverse velocity (v_t):

$$\text{KE}_{\text{segment}} = \frac{1}{2}mv_t^2.$$

The mass of a string segment with a linear mass density of μ and a length of dx is $m = \mu\, dx$, so

$$\text{KE}_{\text{segment}} = \frac{1}{2}(\mu\, dx)v_t^2.$$

The transverse velocity of the segment is $v_t = \partial y/\partial t$, so

$$\text{KE}_{\text{segment}} = \frac{1}{2}(\mu\, dx)\left(\frac{\partial y}{\partial t}\right)^2. \tag{4.15}$$

In the case of harmonic waves, this means that the kinetic energy is greatest for those segments that are passing through the equilibrium position, as you can see by comparing the graphs of $y(x, t)$ and v_t in Fig. 4.3. And, as you might expect, the kinetic energy is zero for segments that are at maximum displacement from equilibrium, since those segments have (instantaneously) zero velocity.

One thing worth noting is that, by setting the length of the segment to the horizontal distance dx, we haven't accounted for the stretching of the string as the segment is displaced from the equilibrium position. This works when you're dealing with the segment's kinetic energy, because it's really the mass of the segment that matters, and any increase in the length of the segment will be accompanied by a proportional decrease in its linear mass density. But, to determine the potential energy of the segment, it's essential that we keep track of the stretching, because the potential energy is related to the work done by the tension forces that cause the segment to stretch.

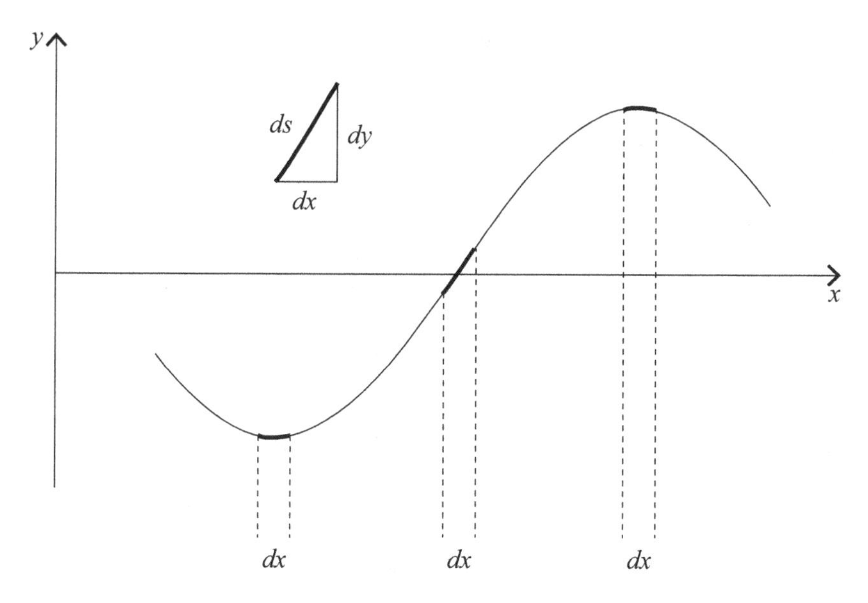

Figure 4.6 Stretching of segments of a harmonic string wave.

To find out how much work is done by those tension forces, the first step is to determine the amount of stretching that occurs as the wave displaces segments of the string from their equilibrium positions. A sketch of the situation is shown in Fig. 4.6, and in that figure you can see that the length of the segment depends on the slope of the string. Approximating the segment length (ds) as the hypotenuse of the right triangle with horizontal side dx and vertical side dy allows you to write that length as

$$ds = \sqrt{dx^2 + dy^2}.$$

If you let dx approach zero, the vertical extent of the segment (dy) can be written as

$$dy = \frac{\partial y}{\partial x}\,dx,$$

so the segment length approaches

$$\begin{aligned} ds &= \sqrt{dx^2 + \left(\frac{\partial y}{\partial x}\,dx\right)^2} \\ &= dx\sqrt{1 + \left(\frac{\partial y}{\partial x}\right)^2}. \end{aligned}$$

This expression can be simplified using the binomial theorem, which says that

$$(1+x)^n \approx 1 + nx$$

as long as x is small relative to one. This approximation works for small-slope wave motion, so you can write

$$\begin{aligned} ds &= dx\left[1+\left(\frac{\partial y}{\partial x}\right)^2\right]^{1/2} \\ &\approx dx\left[1+\frac{1}{2}\left(\frac{\partial y}{\partial x}\right)^2\right] \\ &\approx dx + \frac{1}{2}\left(\frac{\partial y}{\partial x}\right)^2 dx. \end{aligned}$$

This means that the segment is stretched by an amount $ds - dx$, which is

$$\text{Amount of stretch} = ds - dx = \frac{1}{2}\left(\frac{\partial y}{\partial x}\right)^2 dx.$$

To find the work done by the elastic (tension) force in stretching the string by this amount, you can utilize the fact that the work done by any elastic force in stretching an object by some amount equals the component of the force in the direction of the stretch multiplied by the amount of stretch. In this case the elastic force is the tension (T) of the string, so the work is

$$\text{Work} = T\left[\frac{1}{2}\left(\frac{\partial y}{\partial x}\right)^2 dx\right].$$

This work is the change in the potential energy (PE) of the segment, so, if you define the unstretched segment to have zero potential energy, this is the potential energy of the segment:

$$\text{PE}_{\text{segment}} = T\left[\frac{1}{2}\left(\frac{\partial y}{\partial x}\right)^2 dx\right]. \tag{4.16}$$

The total mechanical energy (ME) of any segment of the string is the sum of the segment's kinetic and potential energies, which is

$$\text{ME}_{\text{segment}} = \frac{1}{2}(\mu\, dx)\left(\frac{\partial y}{\partial t}\right)^2 + T\left[\frac{1}{2}\left(\frac{\partial y}{\partial x}\right)^2 dx\right].$$

This is the mechanical energy contained in length dx of the string, so the mechanical energy density (the energy per unit length) can be found by dividing this expression by dx (the horizontal segment length):

$$\mathrm{ME}_{\text{unit length}} = \frac{1}{2}\mu\left(\frac{\partial y}{\partial t}\right)^2 + T\left[\frac{1}{2}\left(\frac{\partial y}{\partial x}\right)^2\right]. \tag{4.17}$$

When you read about mechanical waves in physics texts, you may encounter an expression for energy density written in terms of the wave phase speed (v_{phase}) and the transverse velocity of the string (v_{t}). To see how that comes about, remember that for any wave with wavefunction $y(x, t) = f(x - v_{\text{phase}}t)$ (that is, traveling in the positive x-direction), the time and space derivatives of the wavefunction are related by

$$\frac{\partial y}{\partial x} = \frac{-1}{v_{\text{phase}}}\frac{\partial y}{\partial t}, \tag{4.18}$$

which means the mechanical energy density is

$$\begin{aligned}\mathrm{ME}_{\text{unit length}} &= \frac{1}{2}\mu\left(\frac{\partial y}{\partial t}\right)^2 + T\left[\frac{1}{2}\left(\frac{-1}{v_{\text{phase}}}\frac{\partial y}{\partial t}\right)^2\right]\\ &= \frac{1}{2}\left(\mu + \frac{T}{v_{\text{phase}}^2}\right)\left(\frac{\partial y}{\partial t}\right)^2.\end{aligned}$$

But the phase speed is related to the string's tension and linear mass density by $v_{\text{phase}} = \sqrt{T/\mu}$, which means that $\mu = T/v_{\text{phase}}^2$. Plugging this expression for μ into the previous equation for the energy density gives

$$\begin{aligned}\mathrm{ME}_{\text{unit length}} &= \frac{1}{2}\left(\frac{T}{v_{\text{phase}}^2} + \frac{T}{v_{\text{phase}}^2}\right)\left(\frac{\partial y}{\partial t}\right)^2\\ &= \left(\frac{T}{v_{\text{phase}}^2}\right)\left(\frac{\partial y}{\partial t}\right)^2.\end{aligned}$$

Now just use the fact that $\partial y/\partial t$ is the transverse speed v_{t}, and you have

$$\mathrm{ME}_{\text{unit length}} = \left(\frac{T}{v_{\text{phase}}^2}\right)v_{\text{t}}^2. \tag{4.19}$$

This expression for the string's energy density is helpful if you wish to determine the power flow of a transverse wave on a string, but before doing that it's worthwhile to take a few minutes to apply these general relations to the specific case of a harmonic wave, which is done in the following example.

Example 4.3 *What are the kinetic, potential, and total mechanical energy of a segment of string of length dx with wavefunction* $y(x, t) = A\sin(kx - \omega t)$*?*

For this wavefunction, the transverse velocity is given by $v_t = \partial y/\partial t = -A\omega\cos(kx - \omega t)$, so by Eq. (4.15) the kinetic energy (KE) is

$$\mathrm{KE_{segment}} = \frac{1}{2}(\mu\, dx)\left(\frac{\partial y}{\partial t}\right)^2 = \frac{1}{2}\mu A^2\omega^2\cos^2(kx - \omega t)dx \tag{4.20}$$

and the slope of the wavefunction is $\partial y/\partial x = Ak\cos(kx - \omega t)$, so by Eq. (4.16) the potential energy (PE) is

$$\mathrm{PE_{segment}} = T\left[\frac{1}{2}\left(\frac{\partial y}{\partial x}\right)^2 dx\right] = T\left[\frac{1}{2}A^2k^2\cos^2(kx - \omega t)dx\right].$$

The tension (T) can be eliminated from this equation using the relationships $v_{\mathrm{phase}} = \sqrt{T/\mu}$ and $v_{\mathrm{phase}} = \omega/k$, which can be combined to give $T = \mu\omega^2/k^2$. Thus

$$\mathrm{PE_{segment}} = \left(\mu\frac{\omega^2}{k^2}\right)\frac{1}{2}A^2k^2\cos^2(kx - \omega t)dx$$

or

$$\mathrm{PE_{segment}} = \frac{1}{2}\mu A^2\omega^2\cos^2(kx - \omega t)dx. \tag{4.21}$$

If you compare Eq. (4.21) with Eq. (4.20), you'll see that the segment's kinetic and potential energies are identical. Adding these expressions together gives the total energy density:

$$\mathrm{ME_{segment}} = \mu A^2\omega^2\cos^2(kx - \omega t)dx. \tag{4.22}$$

Equation (4.21) can help you understand why the characterization of potential energy as "energy of position" can be misleading in the case of a transverse wave on a string. For many students, that characterization implies that the potential energy will be greatest for the segment at greatest distance from equilibrium, which is true in the case of a simple harmonic oscillator such as a mass on a spring. But, in the case of a transverse wave on a string, it's not the position of the segment relative to its equilibrium (horizontal) position that determines the segment's potential energy, but the *length* of the segment relative to the equilibrium length. That stretched length depends on the slope ($\partial y/\partial x$) of the segment, which you can see in Fig. 4.6 to be greatest for segments that are passing through the equilibrium position. The segments at maximum displacement are essentially horizontal and unstretched by the wave, which means they have zero potential energy. You can see the total energy of the string segments as a function of x in Fig. 4.7.

It's important for you to realize that this means that the mechanical energy of a segment does not oscillate between all kinetic and all potential, as it

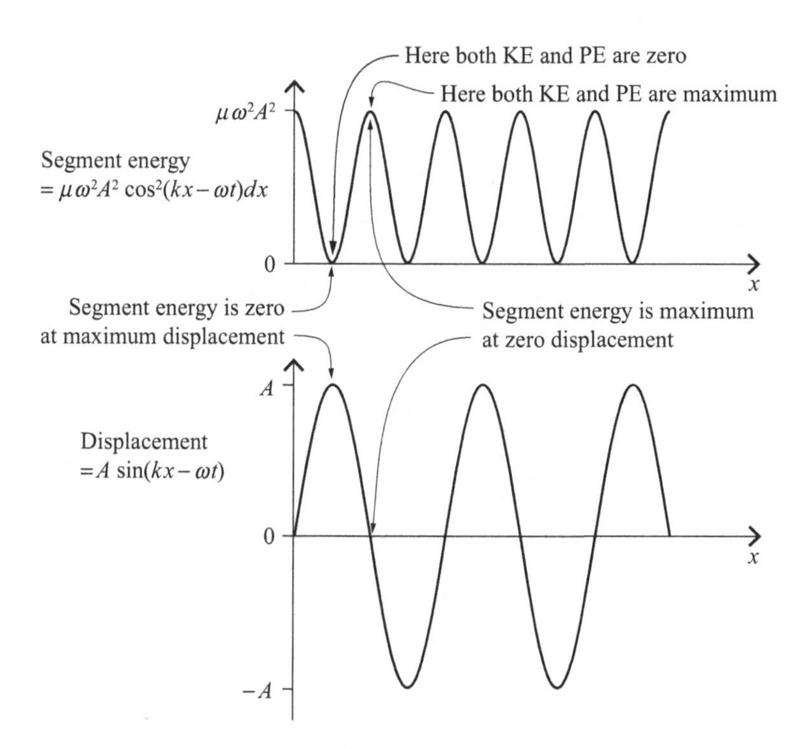

Figure 4.7 Harmonic string wave energy.

does for a mass on a spring undergoing simple harmonic motion. In the case of a transverse string wave, the segment's kinetic and potential energy both reach their maximum value at the same time, which occurs when the segment passes through the equilibrium position. If you're worried that this violates the conservation of energy, just remember that at the instant that some segments have maximum kinetic and potential energy, other segments have zero energy, so the total energy remains constant as long as no dissipative forces are acting.

To find the energy contained not just in a segment of horizontal extent dx but in an entire wavelength of the wave, you can integrate Eq. (4.22) over a distance of one wavelength (λ):

$$\text{ME}_{\text{one wavelength}} = \int_0^\lambda \mu A^2\omega^2 \cos^2(kx - \omega t)dx,$$

which can be done by selecting a fixed time such as $t = 0$ and using the definite integral

$$\int_0^\lambda \cos^2\left(\frac{2\pi}{\lambda}x\right)dx = \frac{\lambda}{2}.$$

Thus the mechanical energy in each wavelength of a transverse string wave is

$$\text{ME}_{\text{one wavelength}} = \frac{1}{2}\mu A^2\omega^2\lambda.$$

Notice that the mechanical energy is proportional to the square of the maximum displacement (A). Although we derived this result using transverse waves on a string, it applies to other forms of mechanical waves, as well. For example, the energy in a pressure wave is proportional to the square of the maximum overpressure of the wave.

With expressions for the energy density and phase speed of a mechanical wave in hand, you can readily find the power of the wave. Power is defined as the rate of change of energy and therefore has SI units of joules per second or watts, and the power of a propagating wave tells you the amount of energy that passes a given location per unit time. Since the mechanical energy density ($\text{ME}_{\text{unit length}}$) is the number of joules in the wave per meter of distance and the phase speed (v_{phase}) is the number of meters the wave moves per second, the product of these two quantities gives the power of the wave:

$$P = (\text{ME}_{\text{unit length}})v_{\text{phase}}.$$

The mechanical energy density is given in Eq. (4.19), so

$$\begin{aligned} P &= \left[\left(\frac{T}{v_{\text{phase}}^2}\right)v_{\text{t}}^2\right]v_{\text{phase}} \\ &= \left(\frac{T}{v_{\text{phase}}}\right)v_{\text{t}}^2. \end{aligned}$$

Since $v_{\text{phase}} = \sqrt{T/\mu}$, the power of the wave is

$$P = \left(\frac{T}{\sqrt{T/\mu}}\right)v_{\text{t}}^2$$

or

$$P = (\sqrt{\mu T})v_{\text{t}}^2. \tag{4.23}$$

The quantity $\sqrt{\mu T}$ in this equation is very important, because it represents the "impedance" of the medium in which the wave propagates (usually denoted by Z). You can read about the physical significance of impedance and its role in the transmission and reflection of waves in Section 4.5, but first here's a short example about the power in a harmonic mechanical wave.

Example 4.4 *Find the power in a transverse mechanical wave with wave-function* $y(x,t) = A\sin(kx - \omega t)$.

As discussed earlier in this section, for this type of harmonic wave the transverse velocity is $v_{\text{phase}} = \sqrt{T/\mu}$, which can be combined with the expression $v_{\text{phase}} = \omega/k$ to give $T = \mu\omega^2/k^2$. Thus

$$P = (\sqrt{\mu T})v_t^2 = \left[\sqrt{\mu\left(\mu\frac{\omega^2}{k^2}\right)}\right]v_t^2$$
$$= \mu\frac{\omega}{k}v_t^2.$$

But for this wave $v_t = -\omega A\cos(kx - \omega t)$, so

$$P = \mu\frac{\omega}{k}[-\omega A\cos(kx - \omega t)]^2$$
$$= \mu\frac{\omega^3}{k}[A^2\cos^2(kx - \omega t)].$$

To find the average power, recall that the average value of $\cos^2$ over many cycles is 1/2, so the average power in a harmonic wave of amplitude A is

$$P_{\text{avg}} = \mu\frac{\omega^3}{k}\left[A^2\left(\frac{1}{2}\right)\right] = \frac{1}{2}\mu A^2\omega^2\left(\frac{\omega}{k}\right)$$

or

$$P_{\text{avg}} = \frac{1}{2}\mu A^2\omega^2 v_{\text{phase}} = \frac{1}{2}ZA^2\omega^2. \tag{4.24}$$

We've written this in several forms to emphasize the relationships between the string parameters of linear mass density μ, phase velocity v_{phase}, and impedance Z (which is $\sqrt{\mu T} = \sqrt{\mu^2\omega^2/k^2} = \mu v_{\text{phase}}$).

4.5 Wave impedance, reflection, and transmission

Although a term representing impedance fell directly out of the derivation of the power of a mechanical wave in the previous section, the physical meaning of impedance is a little more apparent if you come at it from another direction. You'll see that direction at the start of this section, after which you can read about the ways impedance can be used to determine what happens when a wave propagates between two different media.

To understand the physical significance of impedance, consider the force that must be exerted by the source of a mechanical wave in order to produce the initial displacement of the material of the medium. In the case of a transverse mechanical wave, the source (which might be your hand moving one end of a string vertically) must overcome the vertical component of the tension force.

If the displacement of the string from the equilibrium (horizontal) position is y and if the angle of the string with respect to the horizontal is θ, the vertical component of the tension force is

$$F_y = T\sin\theta \approx T\frac{\partial y}{\partial x}.$$

For any single traveling wave, you can use Eq. (4.18) to write this in terms of $\partial y/\partial t$:

$$F_y = T\left(\frac{-1}{v_{\text{phase}}}\right)\frac{\partial y}{\partial t}$$

and, since $\partial y/\partial t = v_{\text{t}}$, this is

$$F_y = T\left(\frac{-1}{v_{\text{phase}}}\right)v_{\text{t}} = -\left(\frac{T}{v_{\text{phase}}}\right)v_{\text{t}}$$

or, since $v_{\text{phase}} = \sqrt{T/\mu}$,

$$F_y = -\left(\frac{T}{\sqrt{T/\mu}}\right)v_{\text{t}} = -(\sqrt{\mu T})v_{\text{t}}. \tag{4.25}$$

This is the force with which the string opposes the motion of the source. The source must overcome this "drag" force by producing a force $F_{y,\text{source}}$ in the opposite direction. Thus the force needed to generate the wave is proportional to the transverse velocity of the string, a fact that will be important when we develop a model of the interaction between a wave and the medium in which it's propagating.

Another important aspect of Eq. (4.25) is that the constant of proportionality between force and transverse velocity is the impedance (Z) of the medium. Notice that, for transverse waves on a string, the impedance depends on only two characteristics of the string: the tension and the linear mass density, $Z = \sqrt{\mu T}$.

Rearranging Eq. (4.25) and using $F_{y,\text{source}} = -F_y$ helps make the meaning of impedance clearer:

$$Z = \sqrt{\mu T} = \frac{F_{y,\text{source}}}{v_{\text{t}}}. \tag{4.26}$$

So, for a mechanical wave, the impedance tells you the amount of force necessary to produce a given transverse velocity of the material displaced by the wave. In SI units, Z gives the number of newtons needed to cause the material to move at a speed of one meter per second. For transverse waves on a string, if the product of the tension and the linear mass density is bigger in one string than it is in another, that string has higher impedance, so it takes more force to achieve a given transverse velocity in that string. But once you succeed in achieving that transverse velocity, the higher-impedance string has more power than a lower-impedance string with the same transverse velocity, as you can see in Eq. (4.23).

Thus the impedance of a medium is very useful if you want to determine the force required to produce a mechanical wave with a certain transverse velocity and if you want to calculate the power of a mechanical wave. But the real power of impedance comes about when you consider what happens to a wave that impinges onto a boundary between two media with different characteristics.

The analysis of the effect of changing media on a mechanical wave begins with the concepts underlying Eq. (4.25). Those concepts are the following.

(1) The medium (the string in this case) produces a drag force on the source of the wave.
(2) That drag force is proportional to the transverse velocity produced in the medium by the wave, and in the opposite direction (so $F_y \propto -v_t$).
(3) The constant of proportionality between the force and the transverse velocity is the impedance (Z) of the medium.

These concepts are very useful for analyzing the behavior of real strings, which don't extend to infinity; if the wave source is at the left end of the string, the right end is some finite distance away. At the right end, the string might be clamped to a wall, it might be free, or it might be attached to another string with different characteristics (such as μ or T, which generally mean different v_{phase} and Z). These situations are shown in Fig. 4.8, and, if you want to know what happens to a rightward-propagating wave (such as the pulse shown in the figure) when it encounters the right end of the string, it's very helpful to understand the nature of the force at the right end of the string (that is, the force that would have been produced by the remainder of the infinite string).

To do that, consider this: Since the drag force produced by the string is proportional to $-v_t$, which varies over space and time, the peaks of the drag force line up with the valleys of v_t. One way to describe this is to say that the drag force is 180° out of phase with the transverse velocity.[4] But, as described previously and shown in Fig. 4.3, the peaks and valleys of the displacement y and transverse acceleration a_t are shifted relative to v_t; they line up with the zero-crossings of v_t. Since the drag force is proportional to $-v_t$, the zero-crossings of v_t are also the zero-crossings of the drag force. Hence the displacement y and the transverse acceleration a_t are ±90° out of phase with the drag force. To see why this matters, take a look at the situations shown in Fig. 4.9.

[4] Another way to say the same thing is that the drag force is in phase with $-v_t$.

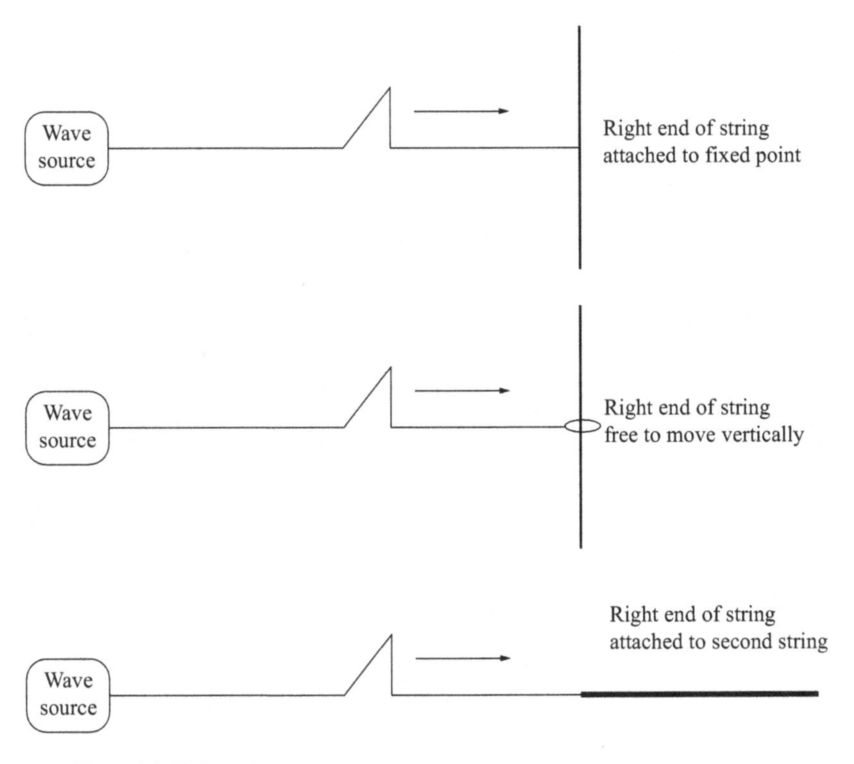

Figure 4.8 Finite strings.

The top portion of this figure shows an infinite string, and that string presents a drag force of $-Zv_t$ to the source of the waves. In that situation, the wave produced by the source propagates continually to the right, and no additional waves are produced (that is, if the string is continuous and has the same properties all along its length, no reflected waves are produced).

Now consider what might happen if you were to eliminate the right portion of the string (that is, imagine that the dashed portion of the string in the lower three sketches in Fig. 4.9 is missing). Clearly the wave will not propagate into the dashed region, because in that region there's no medium to carry the wave, and mechanical waves always require a medium in which to propagate. If you just take away the right portion of the wave, you wouldn't expect the wave in the left portion to behave in the same way it did when the string extended to infinity, since the missing piece is no longer contributing to the drag force.

But a very worthwhile question to ask is this: What could you attach to the string at its right end that would produce a drag force equal to that of

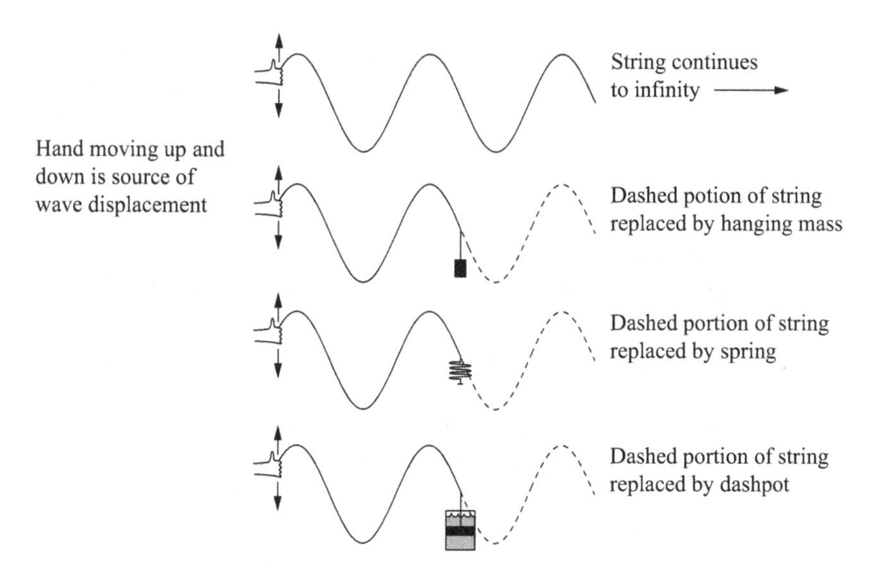

Figure 4.9 Impedance models.

the missing portion of the string? If you could do that, a wave on the left portion would behave exactly as it would on an infinite string. For example, you might try attaching a mass on the end of the string to make up for the mass of the (missing) right portion of the string, as shown in the second sketch from the top. The problem with this approach is that the force produced by the attached mass would not be proportional to the transverse velocity of the incoming wave; instead, it would be proportional to the transverse acceleration (since Newton's second law says that force is proportional to acceleration). And, as shown in Fig. 4.3, the transverse acceleration is zero at the location at which the transverse velocity is maximum. So the drag force would be different, and the wave impinging from the left portion of the figure onto the hanging mass would not behave the same as it would if the string extended to infinity.

As an alternative to attaching a mass to the right end of the string, you might decide to attach a spring, as shown in the third sketch from the top of Fig. 4.9. But once again, the force produced by the spring would not be proportional to the transverse velocity of the incoming wave. In this case it would be proportional to the displacement (since Hooke's law says that the force of a spring is proportional to the displacement from the equilibrium position). And, just like acceleration, displacement is zero at the location at which the transverse velocity is maximum. So a wave on the left portion of the string would not behave the same as it would if the string extended to infinity.

Now imagine attaching a device called a "dashpot" to the right end of the string. If you look up the definition of a dashpot, you're likely to find something like this: "(A dashpot is) a mechanical device which resists motion using viscous friction (such as a piston immersed in a fluid), for which the resulting force is proportional to the velocity, but acts in the opposite direction." That's exactly what you're looking for! On replacing the missing portion of the infinite string with a dashpot, the drag force on the left portion of the string has the same dependence on v_t as it would if the string extended to infinity. And, if you adjust the *amount* of drag produced by the dashpot to exactly match the amount of drag produced by the missing portion of the infinite string, a wave on the left portion of the string will behave exactly as it would if the string extended to infinity (you can think of that as adjusting the impedance of the dashpot to match the impedance of the string). That means that all of the energy of the wave will go into the dashpot, just as it would go into the right portion of an infinite string.

In some texts, you may see the dashpot described as a "purely resistive" device. You should take a few minutes to understand that terminology, because it's based in a very useful parallel between mechanical waves and the time-varying current in an electrical circuit. The reason a dashpot is called a purely resistive device is that the dashpot drag force on the left end of the string is matched in phase to $-v_t$, just as the current in a resistor in an electrical circuit is in phase with the applied voltage. If you've studied alternating-current (AC) electrical circuits, you may recall that the current in other electrical devices, such as capacitors and inductors, is shifted in phase by $\pm 90°$ from the voltage. This is analogous to the phase relationships of the forces produced by the hanging mass and spring discussed above.

So here's the important point: If you terminate a finite-length string with a dashpot and you match the impedance of the dashpot to the impedance of the string, the dashpot acts as a perfect absorber of the wave energy. That means that the drag force produced by the dashpot has just the right amplitude and just the right phase to make the left part of the wave behave as though the string were infinitely long. That behavior is for the wave to propagate to the right without producing any reflected wave.

But how would the wave behave if you adjusted the dashpot's impedance to be somewhat bigger or smaller than the string's impedance? In that case, the drag force would still be proportional to $-v_t$, but the magnitude of the force would be different from the drag force produced by the missing right end of the infinite string. That would mean that some (but not all) of the wave's energy would be absorbed by the dashpot, and some of the wave's energy would be reflected back along the left portion of the string. This mirrors the behavior of

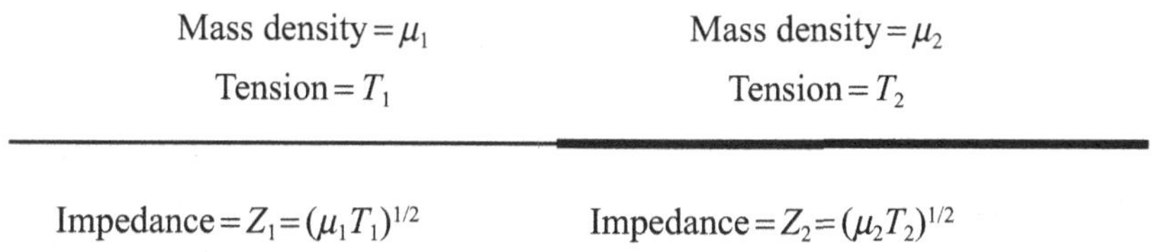

Mass density $= \mu_1$ Tension $= T_1$

Mass density $= \mu_2$ Tension $= T_2$

Impedance $= Z_1 = (\mu_1 T_1)^{1/2}$

Impedance $= Z_2 = (\mu_2 T_2)^{1/2}$

Figure 4.10 An interface between different strings.

a wave that is partially reflected and partially transmitted through an interface. So you can use the dashpot model to investigate the behavior of the wave when the right end of the string is terminated in various ways.

For example, if you clamp the right end of the string to a fixed point, that's like hooking the string to a dashpot with very high impedance, but, if you leave the right end of the string free, that's the equivalent of attaching a dashpot with zero impedance. Now imagine hooking the right end of the string to another string, as shown in Fig. 4.10. If the second string has different values of linear mass density (μ) or tension (T), and thus different impedance (call it Z_2), you can determine how the wave will behave by considering what would happen if you attached the string to a dashpot with impedance (Z_2).

And how would the wave behave under these different circumstances (that is, with different values of Z terminating the string)? To answer that question, it's necessary to enforce two boundary conditions at the interface (the point of termination of the left string). Those boundary conditions are that

(1) the string is continuous, so the displacement (y) must be the same on either side of the interface (if it weren't, the string would be broken at the interface); and
(2) the tangential force ($-T\partial y/\partial x$) must be the same on either side of the interface (if it weren't, a particle of infinitesimal mass at the interface would have almost infinite acceleration).

Applying these boundary conditions leads to the following equation for the displacement (y) of the reflected wave (you can see the details of how this comes about on the book's website):

$$y_{\text{reflected}} = \frac{Z_1 - Z_2}{Z_1 + Z_2} y_{\text{incident}},$$

which tells you that the amplitude of the reflected wave will be bigger or smaller than the amplitude of the incident wave by a factor of $(Z_1 - Z_2)/(Z_1 + Z_2)$. That factor, often denoted r, is called the amplitude reflection coefficient:

$$r = \frac{Z_1 - Z_2}{Z_1 + Z_2}. \tag{4.27}$$

If $r = 1$, the reflected wave has the same amplitude as the incident wave, which means that the wave is totally reflected from the interface – the interface produces an exact copy of the wave that propagates to the left. So in this case the total wave on the string is the superposition of the incident wave and the reflected wave. Alternatively, if $r = 0$, there is no reflected wave, so the only wave on the string is the original rightward-moving wave produced by the source. But, if you look carefully at Eq. (4.27), you'll see that r may also be negative. So, if $r = -1$, for example, the reflected wave has the same amplitude as the incident wave but is inverted. The total wave on the string will be the superposition of the original wave and the inverted counter-propagating wave. These conditions ($r = 1$ and $r = -1$) are the extreme cases; the amplitude reflection coefficient must always have values between these limits.

Note carefully that it's not just the impedance of the second medium that determines the amplitude of the reflected wave – it's the *difference* in impedance between the first medium and the second medium ($Z_1 - Z_2$). So if you want to avoid reflections, you don't try to make the impedance of the second medium as small as possible, you try to match the impedance of the second medium to the impedance of the first medium.

If you wish to know how much of the wave propagates past the interface (the "transmitted wave"), a similar analysis (also on the book's website) shows that

$$y_{\text{transmitted}} = \frac{2Z_1}{Z_1 + Z_2} y_{\text{incident}}.$$

So, calling the amplitude transmission coefficient t (be careful not to confuse this with time), you can write

$$t = \frac{2Z_1}{Z_1 + Z_2}. \tag{4.28}$$

The amplitude transmission coefficient tells you how the amplitude of the transmitted wave compares with the amplitude of the incident wave. If $t = 1$, the transmitted wave has the same amplitude as the incident wave. But, if $t = 0$, the amplitude of the transmitted wave is zero, which means that none of the original wave passes through the interface. For any interface, $t = 1 + r$, so the values of t can range from 0 (if $r = -1$) to +2 (if $r = +1$).

To apply the equations for r and t to a string with the right end clamped to a fixed position, consider that this is equivalent to making $Z_2 = \infty$. In that case,

$$r = \frac{Z_1 - Z_2}{Z_1 + Z_2} = \frac{Z_1 - \infty}{Z_1 + \infty} = -1$$

and

$$t = \frac{2Z_1}{Z_1 + Z_2} = \frac{2Z_1}{Z_1 + \infty} = 0.$$

So in this case none of the incident wave's amplitude is transmitted past the interface, and the reflected wave is an inverted copy of the incident wave.

You can also easily determine what happens in the case in which the right end of the string is left free by setting $Z_2 = 0$; in that case the reflected wave has the same amplitude as the incident wave but is not inverted ($r = 1$).

Knowing how to find r and t from Z_1 and Z_2 also allows you to analyze situations in which a string is attached to another string with different characteristics (mass density μ and tension T). Since $Z = \sqrt{\mu T}$, you simply have to determine the impedance of each string and then use Eq. (4.27) to find r and Eq. (4.28) to find t. Here's an example that illustrates the use of these relationships.

Example 4.5 *Consider a transverse pulse with maximum displacement of 2 cm propagating in the positive x-direction on a string with mass density 0.15 g/cm and tension 10 N. What happens if the pulse enounters a short section of string with twice the mass density and the same tension?*

A sketch of this situation is shown in Fig. 4.11. As you can see in the top portion of the figure, there are two interfaces between the two strings. At the first (left) interface, a pulse traveling in the positive x-direction will be going from a medium with impedance Z_{light} into a medium with impedance Z_{heavy}. So for the rightward-moving pulse at the left interface, $Z_1 = Z_{\text{light}}$ and $Z_2 = Z_{\text{heavy}}$.

As you can see in the lower portion of the figure, some fraction of the pulse will be reflected (leftward) from the left interface, and the remainder of the pulse will be transmitted (rightward) through the first interface. After propagating through the heavy section of string, that transmitted pulse will encounter the second (right) interface. At that interface it will be going from a medium with impedance Z_{heavy} into a medium with impedance Z_{light}. So for a rightward moving pulse at the right interface $Z_1 = Z_{\text{heavy}}$ and $Z_2 = Z_{\text{light}}$. As happened at the left interface, some portion of the pulse will be reflected (leftward) from the second interface, and another portion will be transmitted (rightward) through that interface.

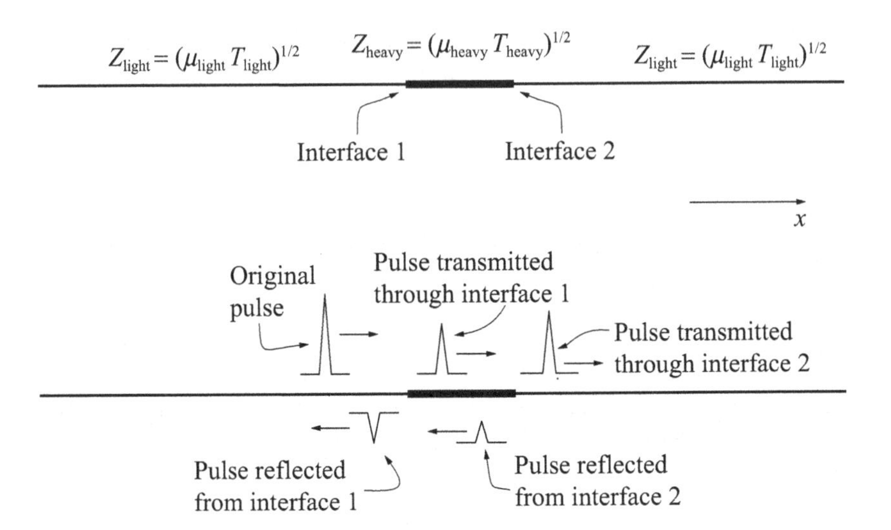

Figure 4.11 A string with a heavy section.

To determine the amplitude of the pulse transmitted through each interface, you can use Eq. (4.28) with the appropriate values of impedances Z_1 and Z_2. You can use Eq. (4.26) to find the impedances after converting the linear mass density to SI units (0.15 g/cm = 0.015 kg/m):

$$Z_1 = \sqrt{\mu_{\text{light}} T_{\text{light}}} = \sqrt{(0.015\,\text{kg/m})(10\,\text{N})} = 0.387\ \text{kg/s},$$
$$Z_2 = \sqrt{\mu_{\text{heavy}} T_{\text{heavy}}} = \sqrt{2(0.015\,\text{kg/m})(10\,\text{N})} = 0.548\ \text{kg/s}.$$

So the transmission coefficient at the left interface is

$$t = \frac{2Z_1}{Z_1 + Z_2} = \frac{(2)(0.387)}{0.387 + 0.548} = 0.83.$$

Thus in propagating from the light string to the heavy string, the amplitude of the pulse is reduced to 83% of its original value. That reduced-amplitude pulse propagates rightward and encounters the second (right) interface. In that case, the heavy string is the medium in which the incident wave propagates, and the light string is the medium of the transmitted wave. Since Z_1 refers to the medium in which the incoming and reflected waves propagate and Z_2 refers to the medium in which the transmitted wave propagates, for this interface the impedances are $Z_1 = 0.548$ kg/s and $Z_2 = 0.387$ kg/s. This makes the transmission coefficient at the right interface

$$t = \frac{2Z_1}{Z_1 + Z_2} = \frac{(2)(0.548)}{0.548 + 0.387} = 1.2,$$

which means that, in propagating past both interfaces, the amplitude of the pulse is reduced by a factor of 0.83 times 1.2, so the final amplitude is about 97% of its original value of 2 cm.

The amplitude reflection and transmission coefficients are useful, but they don't tell you everything about the reflection and transmission process. To see why you need additional information, consider what happens if you subtract the amplitude reflection coefficient r from the amplitude transmission coefficient (t):

$$t - r = \frac{2Z_1}{Z_1 + Z_2} - \frac{Z_1 - Z_2}{Z_1 + Z_2} = \frac{2Z_1 - Z_1 + Z_2}{Z_1 + Z_2}$$
$$= \frac{Z_1 + Z_2}{Z_1 + Z_2} = 1$$

or

$$t = 1 + r.$$

Think about the implications of this equation. In the case of Z_2 much larger than Z_1, the amplitude reflection coefficient approaches -1. It therefore seems reasonable that $t = 1 + r = 1 + (-1) = 0$, since total reflection implies that none of the wave propagates past the interface, so the amplitude of the transmitted wave should be zero.

But what about the case in which Z_1 is much larger than Z_2? For example, what if $Z_2 = 0$? In that case, $r = +1$, and $t = 1 + (+1) = 2$. But if 100% of the incoming wave amplitude is reflected (since $r = 1$), how can the amplitude of the transmitted wave possibly be 2?

To understand the answer to that question, you have to consider the energy being carried by the reflected and transmitted waves, not just their amplitudes. To do that, recall from Eq. (4.24) that the power of the wave is proportional to both the impedance of the medium (Z) and the square of the amplitude (A) of the wave: $P \propto ZA^2$. Thus the ratio of the power in the transmitted wave to the power in the incident wave, called the power transmission coefficient (T, not to be confused with tension) is

$$T = \frac{P_{\text{transmitted}}}{P_{\text{incident}}} = \frac{Z_2 A^2_{\text{transmitted}}}{Z_1 A^2_{\text{incident}}} = \left(\frac{Z_2}{Z_1}\right) t^2, \tag{4.29}$$

since t is the ratio of the transmitted to the incident amplitude. But the reflected wave travels in the same medium (with impedance Z_1) as the incident wave, so the power reflection coefficient (R) is

$$R = \frac{P_{\text{reflected}}}{P_{\text{incident}}} = \frac{Z_1 A^2_{\text{reflected}}}{Z_1 A^2_{\text{incident}}} = \left(\frac{Z_1}{Z_1}\right) r^2 = r^2. \tag{4.30}$$

So the power reflection coefficient (R) is just the square of the amplitude reflection coefficient (r), but the power transmission coefficient (T) is the ratio of the impedances (Z_2/Z_1) times the square of the amplitude transmission coefficient (t).

When you consider the power in the reflected and transmitted waves, it becomes clear that, in the case of $Z_2 = 0$, the power is entirely in the reflected wave (since $R = 1$), with none of the power in the transmitted wave (since $T = 0$).

Since R represents the fraction of the power of the incoming wave that is reflected and T represents the fraction that is transmitted, the sum $R + T$ must equal one (since 100% of the power of the incoming wave must be either reflected or transmitted). You can verify that in one of the chapter-end problems and the online solution.

4.6 Problems

4.1. Show that the expression $\sqrt{T/\mu}$ in Eq. (4.6) has the dimensions of velocity.

4.2. If a string with a length of 2 meters and a mass of 1 gram is tensioned with a hanging mass of 1 kg, what is the phase velocity of transverse waves on the string?

4.3. Show that the expression $\sqrt{K/\rho}$ in Eq. (4.12) has the dimensions of velocity.

4.4. What is the phase speed of pressure waves in an 8-m^3 steel cube, which has a bulk modulus of approximately 150 GPa and a mass of 63,200 kg?

4.5. A transverse harmonic wave with an amplitude of 5 cm and a wavelength of 30 cm propagates on a string of length 70 cm and mass 0.1 gram. If the string is tensioned by a hanging mass of 0.3 kg, what are the kinetic, potential, and total mechanical energy densities of the wave?

4.6. How much power is carried by the wave in the previous problem, and what is the maximum transverse velocity of the string?

4.7. Consider two strings. String A is 20 cm long and has a mass of 12 milligrams, while string B is 30 cm long and has a mass of 25 milligrams. If each string is put under tension by the same hanging mass, how do the speeds of the transverse waves and impedances of the two strings compare?

4.8. If a short segment of the light string of Problem 4.7 is inserted into the heavy string of that problem, find the amplitude reflection coefficients for waves at both interfaces (light-to-heavy and heavy-to-light).

4.9. Find the amplitude transmission coefficients for waves at both interfaces (light-to-heavy and heavy-to-light) in the previous problem.

4.10. Verify that the powers of the transmitted and reflected waves in the previous two problems add up to the power of the incoming wave.

5
The electromagnetic wave equation

This is the second of three chapters that apply the concepts of Chapters 1, 2, and 3 to the three principal types of waves: mechanical, electromagnetic, and quantum. Although the wave nature of the mechanical waves described in Chapter 4 may be more obvious to the casual observer, it is electromagnetic waves that allow us to observe the world around us and the Universe beyond our planet. And, in the last 100 years, we've learned to use electromagnetic waves to send and receive information over distances ranging from a few meters between wireless devices to millions of kilometers between Earth and interplanetary spacecraft.

This chapter begins with an overview of the properties of electromagnetic waves in Section 5.1, after which you'll find a discussion of Maxwell's equations in Section 5.2. The route that leads from those four equations to the electromagnetic wave equation is described in Section 5.3, and plane-wave solutions to that wave equation are described in Section 5.4. This chapter concludes with a discussion of the energy, power, and impedance of electromagnetic waves in Section 5.5.

5.1 Properties of electromagnetic waves

Like any propagating wave, an electromagnetic wave is a disturbance from equilibrium that carries energy from one place to another. But unlike the mechanical waves discussed in Chapter 4, electromagnetic waves do not require a medium in which to propagate.

So if electromagnetic waves can propagate through a vacuum, what exactly is carrying the energy? And what's doing the waving? The answer to both questions is electric and magnetic fields.

Even if you haven't spent much time studying fields, you can still understand the basics of electromagnetic waves. Just remember that fields are closely related to forces; one definition of a field is "a region in which forces act". So an electric field exists in a region in which electric forces act, and a magnetic field exists in a region in which magnetic forces act. That suggests a way you can test for the presence of such fields: Place a charged particle in the region and measure the forces on that particle (your charged particle will have to be moving to detect a magnetic field). But, even if there are no particles to indicate the presence of an electric or magnetic field, there is energy associated with those fields. Thus it's the fields that carry the energy of an electromagnetic wave, and changes in those fields over space and time represent the "waving" of the wave.

To understand the processes at work in electromagnetic waves, it helps to understand how electric and magnetic fields are produced. One source of electric fields is electric charge; as you can see in the left portion of Fig. 5.1, an electric field ($\vec{E}$) diverges from locations of positive charge. All such "electrostatic" fields begin on positive charge and end on negative charge (not shown in the figure). In most diagrams depicting electric and magnetic fields, arrows show the direction of the field at various locations, and the density or length of those arrows indicates the strength of the field. The dimensions of electric fields are force per unit charge, with SI units of newtons per coulomb (N/C), which are equivalent to volts per meter (V/m).

Another source of electric fields is shown in the right portion of Fig. 5.1. That source is a changing magnetic field, and the electric field produced in that

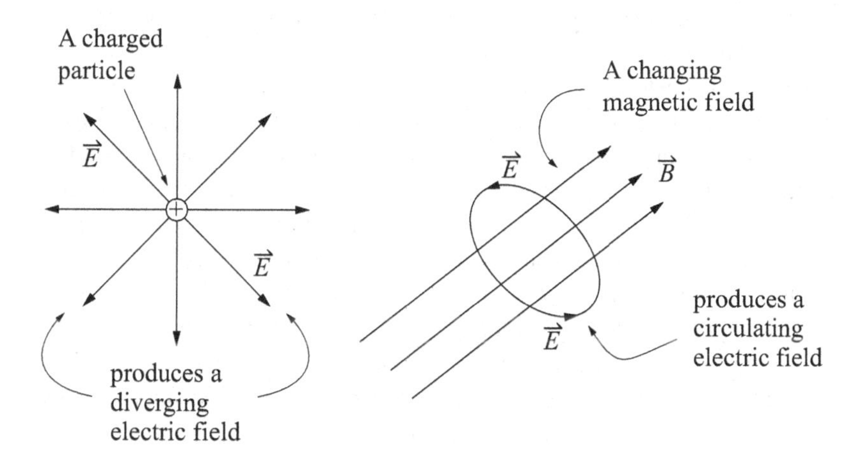

Figure 5.1 Sources of electric fields.

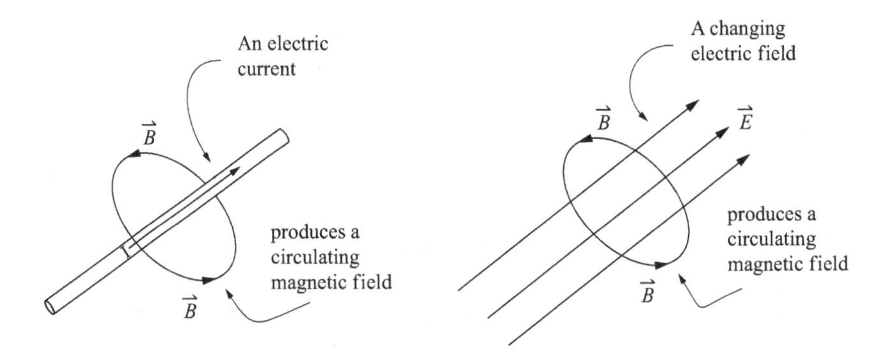

Figure 5.2 Sources of magnetic fields.

case does not diverge from a point, but instead circulates around and connects back upon itself. Such "induced" electric fields are an important component of electromagnetic waves.

Sources of magnetic fields ($\vec{B}$) are shown in Fig. 5.2. Just as electric charges produce an electrostatic field, electric currents produce a "magnetostatic" field that circulates around the current, as shown in the left portion of the figure. And just as a changing magnetic field induces a circulating electric field, a changing electric field induces a circulating magnetic field, as shown in the right portion of the figure. These induced magnetic fields are also an important component of electromagnetic waves. The dimensions of magnetic fields are force divided by current times length, with SI units of newtons per coulomb-meter (N/C-m), which are equivalent to teslas (T).

Since electric and magnetic fields have both magnitude (how strong they are) and direction (which way they point), these fields can be represented by vectors (often signified in text by a little arrow over the symbol representing the vector, such as $\vec{E}$ and $\vec{B}$). If you haven't studied vectors or don't remember much about them, just keep in mind that any vector can be written as a combination of its components (such as E_x, E_y, and E_z or B_x, B_y, and B_z) and basis vectors (such as $\hat{\imath}$, $\hat{\jmath}$, and $\hat{k}$, which point along the x-, y-, and z- axes in the Cartesian coordinate system). So the vector electric field can be written as $\vec{E} = E_x\hat{\imath} + E_y\hat{\jmath} + E_z\hat{k}$, and the vector magnetic field can be written as $\vec{B} = B_x\hat{\imath} + B_y\hat{\jmath} + B_z\hat{k}$ (there's a review of a few basic vector concepts in Section 1.3 of Chapter 1).

The fundamental behavior of vector electric and magnetic fields, and their relationship to one another, are described by four equations known as Maxwell's equations. These equations originated in the work of Gauss, Faraday and Ampère, but it was the brilliant Scottish physicist James Clerk Maxwell who synthesized them and added a critical term to Ampère's law.

As you'll see in this chapter, with that term Maxwell's equations lead directly to the classical wave equation.

The range of applicability of Maxwell's electromagnetic theory is among the greatest in physics. At the low-frequency, long-wavelength end of the electromagnetic spectrum are extremely low-frequency (ELF) radio waves, with frequencies of a few hertz and wavelengths of 100,000 kilometers or longer. At higher frequencies and shorter wavelengths, electromagnetic waves are described as infrared, visible, ultraviolet, X-rays, and gamma rays, spanning a range of some 20 orders of magnitude in frequency and wavelength (so gamma rays have frequencies in excess of 10^{19} Hz and wavelengths smaller than 10^{-11} meters).

In the next section, you can read about each of the four equations that comprise Maxwell's equations.

5.2 Maxwell's equations

Maxwell's equations are four vector equations: Gauss's law for electric fields, Gauss's law for magnetic fields, Faraday's law, and the Ampère–Maxwell law. Each of these equations may be written in integral or differential form. The integral forms describe the behavior of electric and magnetic fields over surfaces or around paths, while the differential forms apply to specific locations. Both forms are relevant to electromagnetic waves, but the trip from Maxwell's equations to the wave equation is somewhat more direct if you start with the differential form. To take advantage of that shorter trip, it's necessary for you to understand the meaning of the vector differential operator called "del" (or "nabla") and written as $\vec{\nabla}$.

A mathematical operator is an instruction to perform an operation, and a differential operator is an instruction to take certain derivatives. So $\sqrt{\ }$ is a mathematical operator telling you to take the square root of whatever appears within the symbol, and $\vec{\nabla}$ is a differential operator telling you to take partial spatial derivatives (such as $\partial/\partial x$, $\partial/\partial y$, and $\partial/\partial z$) of the function on which you're operating. Exactly how those derivatives are to be taken and combined is determined by the symbol that follows the del. So the combination of symbols $\vec{\nabla}\circ$ ("del dot") appearing before vector $\vec{A}$ is defined in Cartesian coordinates as

$$\vec{\nabla} \circ \vec{A} = \frac{\partial A_x}{\partial x} + \frac{\partial A_y}{\partial y} + \frac{\partial A_z}{\partial z}. \tag{5.1}$$

This operation is called "taking the divergence of $\vec{A}$". You can read about the details of divergence in most mathematical physics texts, but the basic idea can be understood through an analogy with flowing fluids.

For this analogy, the vectors of a field (such as $\vec{E}$ or $\vec{B}$) are imagined to represent the flow of some substance toward or away from the location at which the divergence is being taken (although no material is actually flowing in electric or magnetic fields). In this analogy, if the field vectors are such that more material would flow away from a location than toward it, that location has positive divergence. If as much material would flow toward the point as away, the divergence is zero at that location. And if more material would flow toward the point than away, the divergence at that location is negative. To continue the fluid-flow analogy, source points (such as the location at which fluid is flowing from an underwater spring) are locations of positive divergence, and sink points (such as the location at which fluid is flowing down a drain) are locations of negative divergence (sometimes called "convergence").

A helpful thought experiment to judge the divergence at any location in a field is to imagine spreading sawdust on a fluid. That sawdust will disperse if placed at locations of positive divergence, and it will collect in regions of negative divergence.

Applying this concept to electric fields, you can conclude from the left portion of Fig. 5.1 that any location at which positive electric charge exists is a location of positive divergence of the electric field.

Another use of the del operator in Maxwell's equations involves the combination of symbols $\vec{\nabla}\times$ ("del cross"). When these appear before a vector, the operation is defined in Cartesian coordinates as

$$\vec{\nabla}\times\vec{A} = \left(\frac{\partial A_z}{\partial y} - \frac{\partial A_y}{\partial z}\right)\hat{\imath} + \left(\frac{\partial A_x}{\partial z} - \frac{\partial A_z}{\partial x}\right)\hat{\jmath} + \left(\frac{\partial A_y}{\partial x} - \frac{\partial A_x}{\partial y}\right)\hat{k}. \quad (5.2)$$

This operation is called "taking the curl of $\vec{A}$". Again, the basic idea can be understood through an analogy with flowing fluids. In this case the issue is not how much material is being carried toward or away from a point by the vector field, but instead how strongly the field is circulating around the point under consideration. So a point at the center of a swirling vortex is a location of high curl, but smooth flow radially outward from a source has zero curl.

Unlike the divergence, which produces a scalar result (that is, a value with magnitude but no direction), the curl operator produces a vector result. So which way does the curl vector point? It points along the axis of rotation of the field, and, by convention, the direction of positive curl corresponds to the direction your right thumb points if you curl the fingers of your right hand in the direction of the field circulation.

A helpful thought experiment for curl is to imagine placing a small paddlewheel on the end of an axle (a shaft extending from the center of the wheel) into the vector field at the point under consideration. If that location has non-zero curl, the paddlewheel will rotate, and the direction of the curl vector is along the axle (with the positive direction defined by the right-hand rule described above).

Applying this concept to the fields in Fig. 5.2 should help you see that any location at which an electric current or a changing electric field exists is a location of non-zero curl of the magnetic field.

With these concepts in hand, you should be ready to consider the differential form of Maxwell's equations. Here they are, followed by a short description of the physical meaning of each equation.

(I) Gauss's law for electric fields: $\vec{\nabla} \circ \vec{E} = \rho/\epsilon_0$

Gauss's law for electric fields states that the divergence ($\vec{\nabla}\circ$) of the electric field ($\vec{E}$) at any location is proportional to the electric charge density (ρ) at that location. That's because electrostatic field lines begin on positive charge and end on negative charge (hence the field lines tend to diverge away from locations of positive charge and converge toward locations of negative charge). The symbol ϵ_0 represents the electric permittivity of free space, a quantity that you'll see again when we consider the phase speed and impedance of electromagnetic waves.

(II) Gauss's law for magnetic fields: $\vec{\nabla} \circ \vec{B} = 0$

Gauss's law for magnetic fields tells you that the divergence ($\vec{\nabla}\circ$) of the magnetic field ($\vec{B}$) at any location must be zero. This is true because there is apparently no isolated "magnetic charge" in the Universe, so magnetic field lines neither diverge nor converge (they circulate back on themselves).

(III) Faraday's law: $\vec{\nabla} \times \vec{E} = -\partial\vec{B}/\partial t$

Faraday's law indicates that the curl ($\vec{\nabla}\times$) of the electric field ($\vec{E}$) at any location is equal to the negative of the time rate of change of the magnetic field ($\partial\vec{B}/\partial t$) at that location. That's because a changing magnetic field produces a circulating electric field.

(IV) Ampère–Maxwell law: $\vec{\nabla} \times \vec{B} = \mu_0\vec{J} + \mu_0\epsilon_0\partial\vec{E}/\partial t$

Ampère's law, as modified by Maxwell, tells you that the curl ($\vec{\nabla}\times$) of the magnetic field ($\vec{B}$) at any location is proportional to the electric current density ($\vec{J}$) plus the time rate of change of the electric field ($\partial\vec{E}/\partial t$) at that location.[1] This is the case because a circulating magnetic field is produced both by an

[1] The term involving the changing electric field is the "displacement current" added to Ampère's law by James Clerk Maxwell.

electric current and by a changing electric field. The symbol μ_0 represents the magnetic permeability of free space, another quantity that you'll see when we consider the phase speed of electromagnetic waves and the electromagnetic impedance.

Notice that Maxwell's equations relate the spatial behavior of fields to the sources of those fields. Those sources are electric charge (with density ρ) appearing in Gauss's law for electric fields, electric current (with density $\vec{J}$) appearing in the Ampère–Maxwell law, changing magnetic field (with time derivative $\partial\vec{B}/\partial t$) appearing in Faraday's law, and changing electric field (with time derivative $\partial\vec{E}/\partial t$) appearing in the Ampère–Maxwell law.

The next section of this chapter shows you how Maxwell's equations lead to the classical wave equation for electromagnetic waves.

5.3 Electromagnetic wave equation

Taken individually, Maxwell's equations provide important relationships between the sources of electric and magnetic fields and the behavior of those fields. But the real power of these equations is realized by combining them together to produce the wave equation.

To see how that works, start by taking the curl of both sides of Faraday's law:

$$\vec{\nabla} \times (\vec{\nabla} \times \vec{E}) = \vec{\nabla} \times \left(-\frac{\partial \vec{B}}{\partial t}\right)$$
$$= -\frac{\partial(\vec{\nabla} \times \vec{B})}{\partial t},$$

in which the spatial partial derivatives of $\vec{\nabla}\,\times$ have been moved inside the time partial derivative $\partial/\partial t$ (which you're allowed to do for sufficiently smooth functions). Inserting the expression for the curl of the magnetic field ($\vec{\nabla} \times \vec{B}$) from the Ampère–Maxwell law makes this

$$\vec{\nabla} \times (\vec{\nabla} \times \vec{E}) = -\frac{\partial(\mu_0\vec{J} + \mu_0\epsilon_0\, \partial\vec{E}/\partial t)}{\partial t}$$
$$= -\mu_0\frac{\partial \vec{J}}{\partial t} - \mu_0\epsilon_0\frac{\partial^2 \vec{E}}{\partial t^2}.$$

The final steps to the electromagnetic wave equation require the use of the vector identity for the curl of the curl of a function

$$\vec{\nabla} \times (\vec{\nabla} \times \vec{A}) = \vec{\nabla}(\vec{\nabla} \circ \vec{A}) - \nabla^2\vec{A},$$

in which $\vec{\nabla}(\vec{\nabla} \circ \vec{A})$ represents the gradient (the spatial change) of the divergence of $\vec{A}$ and $\nabla^2\vec{A}$ represents the Laplacian of $\vec{A}$, a vector operator involving second-order spatial partial derivatives as described below. Applying this identity to the previous equation gives

$$\vec{\nabla}(\vec{\nabla} \circ \vec{E}) - \nabla^2\vec{E} = -\mu_0\frac{\partial \vec{J}}{\partial t} - \mu_0\epsilon_0\frac{\partial^2 \vec{E}}{\partial t^2}$$

and using Gauss's law for electric fields ($\vec{\nabla} \circ \vec{E} = \rho/\epsilon_0$) makes this

$$\vec{\nabla}\left(\frac{\rho}{\epsilon_0}\right) - \nabla^2\vec{E} = -\mu_0\frac{\partial \vec{J}}{\partial t} - \mu_0\epsilon_0\frac{\partial^2 \vec{E}}{\partial t^2}.$$

In a vacuum, the charge density (ρ) and the current density ($\vec{J}$) are both zero. Thus in free space

$$0 - \nabla^2\vec{E} = 0 - \mu_0\epsilon_0\frac{\partial^2 \vec{E}}{\partial t^2}$$

or

$$\nabla^2\vec{E} = \mu_0\epsilon_0\frac{\partial^2 \vec{E}}{\partial t^2}. \tag{5.3}$$

This is the wave equation for the electric field. Since this is a vector equation, it's actually three separate equations (one for each component of the vector $\vec{E}$). In Cartesian coordinates those equations are

$$\begin{aligned}
\frac{\partial^2 E_x}{\partial x^2} + \frac{\partial^2 E_x}{\partial y^2} + \frac{\partial^2 E_x}{\partial z^2} &= \mu_0\epsilon_0\frac{\partial^2 E_x}{\partial t^2},\\
\frac{\partial^2 E_y}{\partial x^2} + \frac{\partial^2 E_y}{\partial y^2} + \frac{\partial^2 E_y}{\partial z^2} &= \mu_0\epsilon_0\frac{\partial^2 E_y}{\partial t^2},\\
\frac{\partial^2 E_z}{\partial x^2} + \frac{\partial^2 E_z}{\partial y^2} + \frac{\partial^2 E_z}{\partial z^2} &= \mu_0\epsilon_0\frac{\partial^2 E_z}{\partial t^2}.
\end{aligned} \tag{5.4}$$

You can find the equivalent equations for the magnetic field $\vec{B}$ by taking the curl of both sides of the Ampère–Maxwell law and then inserting the curl of $\vec{E}$ from Faraday's law. This gives

$$\nabla^2\vec{B} = \mu_0\epsilon_0\frac{\partial^2 \vec{B}}{\partial t^2}. \tag{5.5}$$

This is also a vector equation, with component equations

$$\frac{\partial^2 B_x}{\partial x^2} + \frac{\partial^2 B_x}{\partial y^2} + \frac{\partial^2 B_x}{\partial z^2} = \mu_0\epsilon_0\frac{\partial^2 B_x}{\partial t^2},$$
$$\frac{\partial^2 B_y}{\partial x^2} + \frac{\partial^2 B_y}{\partial y^2} + \frac{\partial^2 B_y}{\partial z^2} = \mu_0\epsilon_0\frac{\partial^2 B_y}{\partial t^2}, \quad (5.6)$$
$$\frac{\partial^2 B_z}{\partial x^2} + \frac{\partial^2 B_z}{\partial y^2} + \frac{\partial^2 B_z}{\partial z^2} = \mu_0\epsilon_0\frac{\partial^2 B_z}{\partial t^2}.$$

An important aspect of electromagnetic waves can be seen by comparing Eqs. (5.4) and (5.6) with the general equation for a propagating wave (Eq. (2.11) of Chapter 2):

$$\frac{\partial^2 \Psi}{\partial x^2} + \frac{\partial^2 \Psi}{\partial y^2} + \frac{\partial^2 \Psi}{\partial z^2} = \frac{1}{v_{\text{phase}}^2}\frac{\partial^2 \Psi}{\partial t^2}. \quad (2.11)$$

Equating the factors in front of the time derivatives reveals the phase velocity of electromagnetic waves:

$$\frac{1}{v_{\text{phase}}^2} = \mu_0\epsilon_0$$

or

$$v_{\text{phase}} = \sqrt{\frac{1}{\mu_0\epsilon_0}}. \quad (5.7)$$

Hence the velocity of an electromagnetic wave in a vacuum depends only on the electric permittivity (ϵ_0) and magnetic permeability (μ_0) of free space. The values of these constants can be determined experimentally using capacitors and inductors; the accepted values are $\epsilon_0 = 8.8541878 \times 10^{-12}$ farads/meter and $\mu_0 = 4\pi \times 10^{-7}$ henries/meter.[2] Plugging these values into Eq. (5.7) gives

$$v_{\text{phase}} = \sqrt{\frac{1}{\mu_0\epsilon_0}} = \sqrt{\frac{1}{(8.8541878 \times 10^{-12}\ \text{F/m})(4\pi \times 10^{-7}\ \text{H/m})}}$$
$$= 2.9979 \times 10^8\ \text{m/s}.$$

This is the speed of light in a vacuum, an astonishing result that caused Maxwell to conclude that light is an electromagnetic disturbance.

Electromagnetic waves must satisfy not only the wave equation, but also Maxwell's equations. On applying Maxwell's equations to the solutions to the

[2] Farads are units of capacitance, equivalent to (coulombs^2 seconds^2)/(kilograms meters^3), and henries are units of inductance equivalent to (meters kilograms)/coulombs^2.

two separate wave equations (Eqs. (5.3) and (5.5)), the connections between those solutions become clear. You can see how that works in the next section.

5.4 Plane-wave solutions to the electromagnetic wave equation

There are various solutions to the electromagnetic wave equation, and one very important subset of those solutions involves plane waves. In plane waves, the surfaces of constant phase are flat planes perpendicular to the direction of propagation (if you're not sure what that means, it might help to take a quick look ahead to Fig. 5.3). In this section, we'll consider electromagnetic plane waves propagating in the positive z-direction, which means that the planes of constant phase are parallel to the (x, y)-plane.

As explained in Chapter 2, solutions to the wave equation for propagation in the positive z-direction are functions of the form $f(kz - \omega t)$, so one solution for the electric field uses the harmonic function[3]

$$\vec{E} = \vec{E}_0 \sin(kz - \omega t), \tag{5.8}$$

where $\vec{E}_0$ represents the "vector amplitude" of the propagating electric field. As its name implies, the vector amplitude is a vector, which means that it has both magnitude and direction. The magnitude of $\vec{E}_0$ tells you the amplitude of the electric field (that is, the largest value the electric field achieves, which occurs when $\sin(kz - \omega t) = 1$), and the direction of $\vec{E}_0$ tells you which way the electric field points.

Likewise, the solution for the magnetic field can be written as

$$\vec{B} = \vec{B}_0 \sin(kz - \omega t), \tag{5.9}$$

in which $\vec{B}_0$ is the vector amplitude of the propagating magnetic field.

In Cartesian coordinates, the components of the vector amplitudes of the electric and magnetic fields are

$$\vec{E}_0 = E_{0x}\hat{\imath} + E_{0y}\hat{\jmath} + E_{0z}\hat{k} \tag{5.10}$$

and

$$\vec{B}_0 = B_{0x}\hat{\imath} + B_{0y}\hat{\jmath} + B_{0z}\hat{k}. \tag{5.11}$$

[3] If you're concerned that results obtained using harmonic functions may not apply to other waveforms, remember from the discussion of Fourier synthesis in Section 3.3 of Chapter 3 that any well-behaved function can be synthesized from combinations of harmonic functions.

You can learn a lot about these components by applying Maxwell's equations. Starting with Gauss's law for electric fields, you know that $\vec{\nabla} \circ \vec{E} = 0$ in a vacuum, which means

$$\frac{\partial E_x}{\partial x} + \frac{\partial E_y}{\partial y} + \frac{\partial E_z}{\partial z} = 0. \tag{5.12}$$

But, since the phase of the wave must be constant over the entire (x, y)-plane, the electric-field components cannot vary with x or y (that doesn't mean those components are zero, just that each component can't change its value with x or y). Hence the first two derivatives ($\partial E_x/\partial x$ and $\partial E_y/\partial y$) of Eq. (5.12) must be zero, which means that

$$\frac{\partial E_z}{\partial z} = 0.$$

To understand the implication of this, consider the fact that the wave is propagating in the z-direction. So this equation says that, if the electric field has a component in the z-direction, that component must be the same at all values of z (because if E_z changed with z, then $\partial E_z/\partial z$ wouldn't equal zero). But a constant E_z doesn't contribute to the wave disturbance (which is a function of both z and t), so you can take E_z of the wave as zero. Thus a plane-wave electric field has no component in the direction of propagation.

Since Gauss's law for magnetic field says that $\vec{\nabla} \circ \vec{B} = 0$, the same reasoning leads to the conclusion that B_z must also equal zero for a plane wave propagating in the z-direction. And, if $E_z = 0$ and $B_z = 0$, then both the electric field and the magnetic field must be perpendicular to the direction of propagation of the wave. Thus an electromagnetic plane wave is a transverse wave.

The possible components of such a wave are therefore

$$E_x = E_{0x} \sin(kz - \omega t), \qquad B_x = B_{0x} \sin(kz - \omega t),$$
$$E_y = E_{0y} \sin(kz - \omega t), \qquad B_y = B_{0y} \sin(kz - \omega t).$$

Just as Gauss's laws allowed you to eliminate the z-components of the electric and magnetic fields, Faraday's law can be used to gain some insight into the relationship between the remaining components. That equation relates the curl of the induced electric field to the rate of change of the magnetic field over time:

$$\vec{\nabla} \times \vec{E} = -\frac{\partial \vec{B}}{\partial t}.$$

Using the definition of the curl in Cartesian coordinates (Eq. (5.2)), the x-component of this equation is

$$\left(\frac{\partial E_z}{\partial y} - \frac{\partial E_y}{\partial z}\right) = -\frac{\partial B_x}{\partial t},$$

which says that

$$E_{0y} = -cB_{0x}. \tag{5.13}$$

The y-component equation leads to

$$E_{0x} = cB_{0y}. \tag{5.14}$$

If you need some help getting either of these results, see the chapter-end problems and online solutions. These equations reveal the connection between the solutions of the electric-field wave equation and the solutions to the magnetic-field wave equation.

Another important bit of information is contained in Eqs. (5.13) and (5.14). In an electromagnetic plane wave, the electric field and the magnetic field are not only perpendicular to the direction of propagation, but also perpendicular to one another (once again, if you need help seeing why that's true, take a look at the chapter-end problems and online solutions).

Example 5.1 *If an electromagnetic plane wave is propagating along the positive z-direction and its electric field at a certain location points along the positive x-axis, in what direction does the wave's magnetic field point at that location?*

If the electric field points along the positive x-axis, you know that E_{0x} is positive and $E_{0y} = 0$. That means that the magnetic field must have a non-zero component along the positive y-axis, since $B_{0y} = E_{0x}/c$ and E_{0x} is positive. But the fact that $E_{0y} = 0$ means that B_{0x} (which equals $-E_{0y}/c$) must also be zero. And, if B_{0y} is positive and B_{0x} is zero, then $\vec{B}$ must point entirely along the positive y-axis at this location.

A sketch of the electric field of an electromagnetic plane wave propagating in the z-direction is shown in Fig. 5.3. In this case, we've chosen the x-axis to coincide with the direction of the electric field, and the figure shows a snapshot of the wave taken at a time when a positive peak is crossing the (x, y)-plane at location $z = 0$. In the upper portion of the figure, the strength of the electric field is indicated by the density of the arrows, and the lower portion shows a graph of the electric field strength along the z-axis.

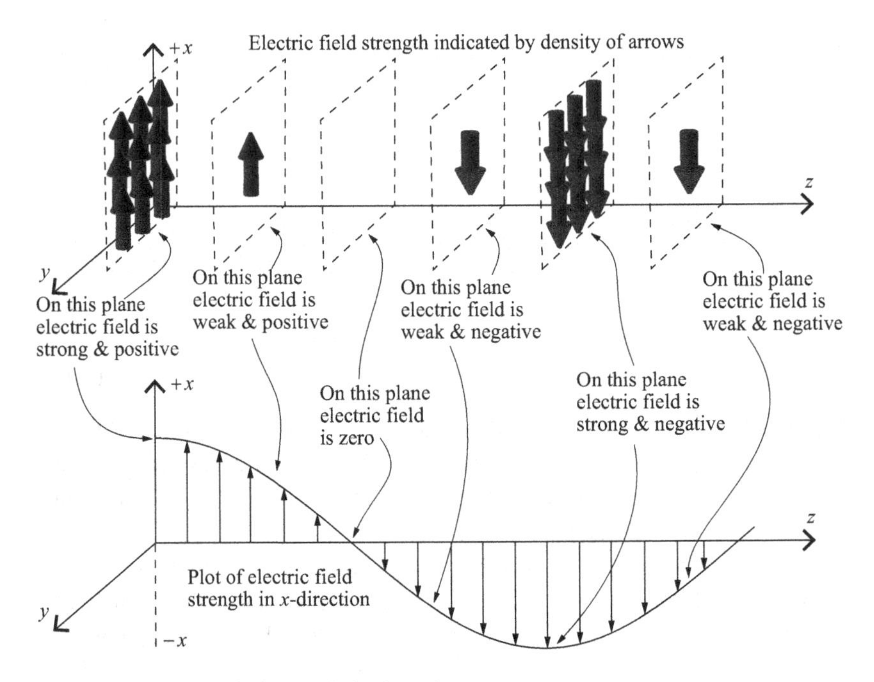

Figure 5.3 Electric field variation in a plane wave.

A corresponding sketch for the magnetic field at the same instant in time is shown in Fig. 5.4. As you can see in this figure, the magnetic field points along the y-axis, perpendicular to both the direction of propagation and the electric field in Fig. 5.3. The plot of the magnetic field strength in the lower portion of the figure lies along the positive and negative y-axis.

Combining the field-strength plots from Figs. 5.3 and 5.4 results in the plot shown in Fig. 5.5. In this figure, the plot of the magnetic field has been scaled to have the same amplitude as the electric field for clarity, but the actual relative magnitudes of the field strengths can be found as

$$|\vec{E}| = \sqrt{(E_{0x})^2 + (E_{0y})^2} = \sqrt{(cB_{0y})^2 + (-cB_{0x})^2}$$
$$= c\sqrt{(B_{0y})^2 + (B_{0x})^2} = c|\vec{B}|$$

or

$$\frac{|\vec{E}|}{|\vec{B}|} = c, \tag{5.15}$$

which says that the electric field strength (in volts per meter) is bigger than the magnetic field strength (in teslas) by a factor of c (in meters per second).

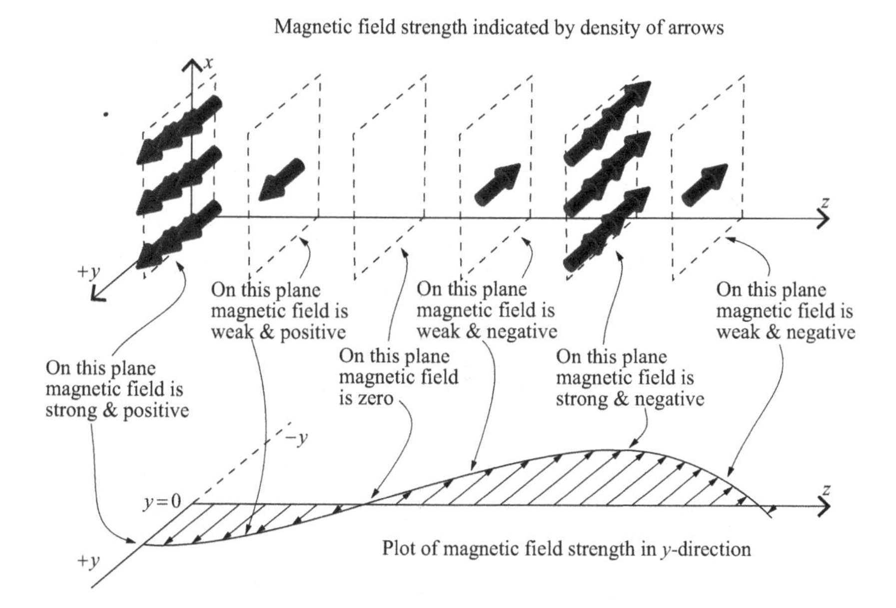

Figure 5.4 Magnetic field variation in a plane wave.

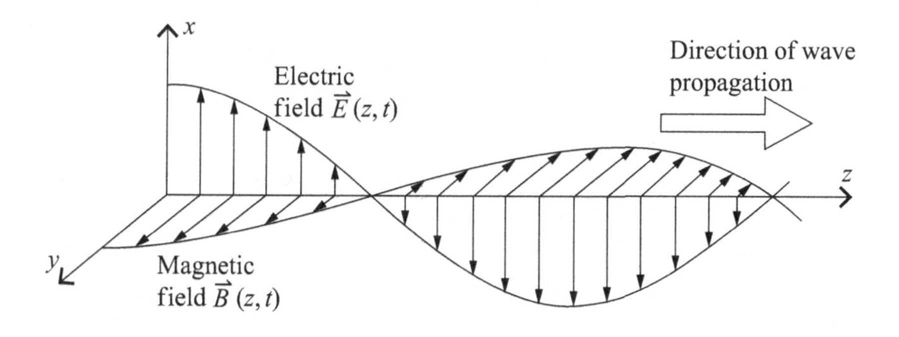

Figure 5.5 Electromagnetic plane wave.

The direction of the electric field defines the "polarization" of the wave; if the electric field remains in the same plane (as it does in this case), the wave is said to be linearly polarized or plane polarized. With an understanding of the behavior of electric and magnetic fields in transverse plane waves, you can also understand more complex electromagnetic waves. For example, consider the plot of the radiation from an oscillating electric dipole shown in Fig. 5.6. Plots like this are often called "radiation patterns", and in these plots the

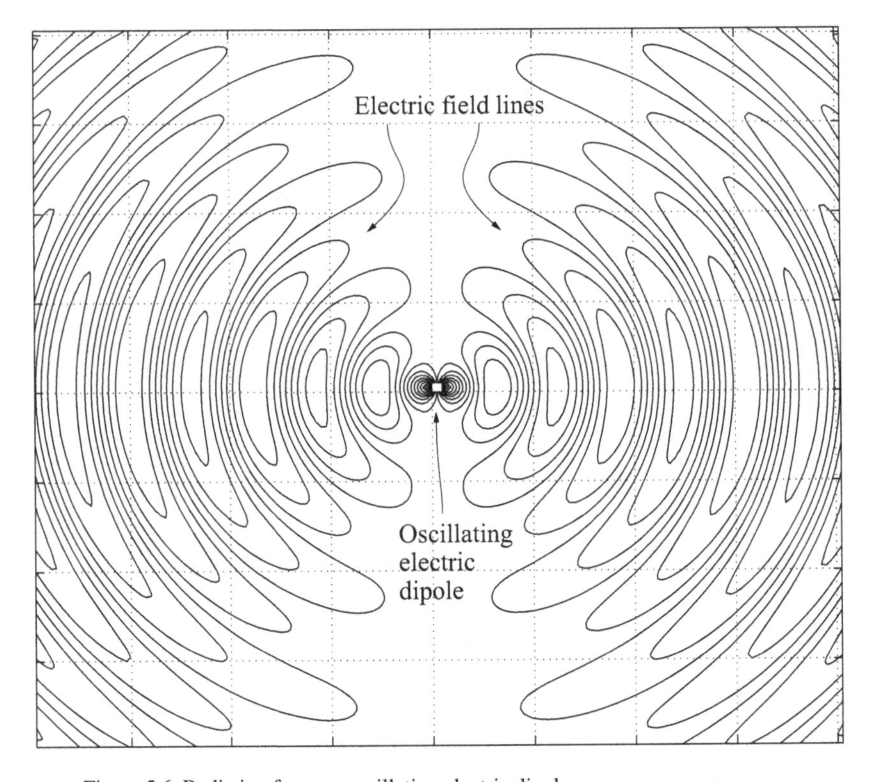

Figure 5.6 Radiation from an oscillating electric dipole.

curving lines represent the electric-field component of the electromagnetic wave produced by the dipole. This plot is a two-dimensional slice through the three-dimensional radiation pattern, so, to visualize the entire field, you should imagine rotating the pattern around a vertical axis passing through the dipole.

A key point is this: Far from the dipole (that is, at distances of several wavelengths or more), the electromagnetic waves produced by an oscillating dipole resemble plane waves to a good approximation. That's because the radius of curvature of the field lines gets larger and larger as the wave propagates away from the source, which means the surfaces of constant phase become more and more like flat planes.

What may not be clear in such plots is how the electric-field lines shown in the radiation pattern relate to the electric and magnetic fields shown in the plane wave in Fig. 5.5. To help you understand that relationship, we've inserted a small, properly scaled (in z) version of the plane-wave field diagram into the dipole radiation pattern in Fig. 5.7.

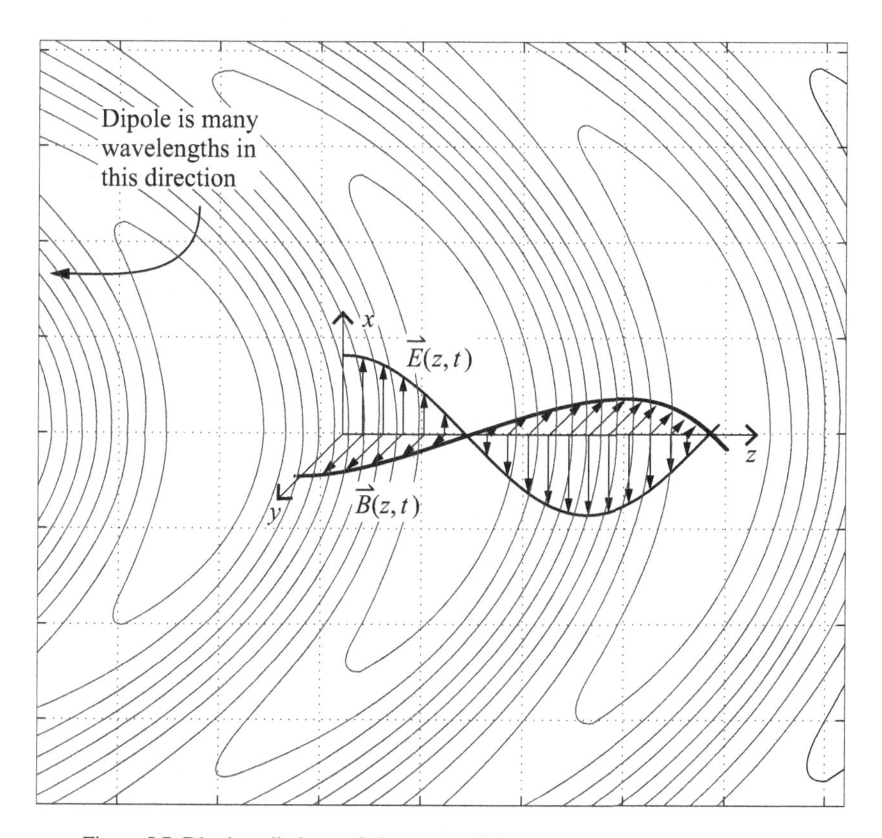

Figure 5.7 Dipole radiation and plane-wave fields.

As you can see in this figure, the electric field is strongest at the locations in the radiation plot at which the electric-field lines are closest together, and the electric field is weakest (passing through zero) at the locations at which the field lines are far apart.

Radiation plots generally don't include the magnetic-field lines, but you can see the direction and strength of the magnetic-field on the graph in the small plane-wave insert in Fig. 5.7. If you imagine rotating the field pattern out of the page and around the dipole, you should be able to see that the magnetic-field lines circulate around the source and close back upon themselves, as Maxwell's equations say they must.

Like all propagating waves, electromagnetic waves carry energy, even when they propagate through a perfect vacuum. This may seem counter-intuitive for a medium with zero mass density, but the energy of an electromagnetic field is contained within the electric and magnetic fields themselves. You can read about electromagnetic energy, power, and impedance in the next section.

5.5 Energy, power, and impedance of electromagnetic waves

To understand the energy stored in electric and magnetic fields, it's helpful to consider the energy stored in a charged capacitor and in a current-carrying inductor. That's because the charge stored on a capacitor produces an electric field between the plates, and the current flowing through an inductor produces a magnetic field. By calculating the amount of work done in establishing those fields, the energy stored in the fields may be determined.[4] If you divide that energy by the volume it occupies, you'll get the energy density stored in the field (that is, the energy per unit volume, with SI units of joules per cubic meter).

For the electric field in a vacuum, the energy density (often designated as u_E) is

$$u_E = \frac{1}{2}\epsilon_0|\vec{E}|^2, \tag{5.16}$$

where ϵ_0 is the electric permittivity of free space and $|\vec{E}|$ is the magnitude of the electric field.

For the magnetic field in a vacuum, the energy density (u_B) is

$$u_B = \frac{1}{2\mu_0}|\vec{B}|^2, \tag{5.17}$$

where μ_0 is the magnetic permeability of free space and $|\vec{B}|$ is the magnitude of the magnetic field.

You should note two things about these relations. First, in both cases the energy density is proportional to the square of the field magnitude, so fields that are twice as strong store four times as much energy. Second, since u_E and u_B are energy densities, if you want to know the energy stored in some region, you have to multiply these quantities by the volume of the region (if the energy density is uniform throughout the volume) or integrate over the volume (if the energy density is a function of position within the volume).

If both electric and magnetic fields are present in some region of space, the total energy density may be found by combining Eqs. (5.16) and (5.17):

$$u_{\text{tot}} = \frac{1}{2}\epsilon_0|\vec{E}|^2 + \frac{1}{2\mu_0}|\vec{B}|^2. \tag{5.18}$$

Another version of this equation that you're likely to encounter uses the ratio of the electric to magnetic field magnitudes to eliminate either $|\vec{E}|$ or $|\vec{B}|$. Since Eq. (5.15) tells you that $|\vec{E}|/|\vec{B}| = c$ in an electromagnetic wave, the total energy may be written as

[4] You can find the details of this in any comprehensive text on electricity and magnetism.

$$u_{\text{tot}} = \frac{1}{2}\epsilon_0|\vec{E}|^2 + \frac{1}{2\mu_0}\left(\frac{|\vec{E}|}{c}\right)^2.$$

But $c = 1/\sqrt{\mu_0\epsilon_0}$, so this is

$$\begin{aligned} u_{\text{tot}} &= \frac{1}{2}\epsilon_0|\vec{E}|^2 + \frac{\mu_0\epsilon_0}{2\mu_0}|\vec{E}|^2 \\ &= \frac{1}{2}\epsilon_0|\vec{E}|^2 + \frac{1}{2}\epsilon_0|\vec{E}|^2 \end{aligned}$$

or

$$u_{\text{tot}} = \epsilon_0|\vec{E}|^2. \tag{5.19}$$

Eliminating $|\vec{E}|$ rather than $|\vec{B}|$ from Eq. (5.18) results in

$$u_{\text{tot}} = \frac{1}{\mu_0}|\vec{B}|^2. \tag{5.20}$$

So either $\vec{E}$ or $\vec{B}$ can tell you the amount of energy per unit volume stored in the fields of an electromagnetic wave.

You also know that this energy is moving at the speed of the wave, which for electromagnetic waves in free space is the vacuum speed of light (c). So consider what happens if you multiply u_{tot} (with SI units of J/m^3) by c (with SI units of m/s): You get joules per second per square meter. That's the rate at which energy flows through a cross-sectional area of one square meter perpendicular to the direction in which the wave is moving. And, since energy per unit time (joules per second in SI units) is power (SI unit of watts), the magnitude of the power per unit area in an electromagnetic wave is

$$|\vec{S}| = u_{\text{tot}}c,$$

in which the power density is written as the magnitude of a vector ($\vec{S}$) called the "Poynting vector" (you can read more about that vector below).

In terms of the electric field, this means that the power density is

$$\begin{aligned} |\vec{S}| &= \epsilon_0|\vec{E}|^2 c = \epsilon_0|\vec{E}|^2\sqrt{\frac{1}{\mu_0\epsilon_0}} \\ &= \sqrt{\frac{\epsilon_0}{\mu_0}}|\vec{E}|^2. \end{aligned} \tag{5.21}$$

So the power density of an electromagnetic plane wave is proportional to the square of the magnitude of the wave's electric field, and the constant of proportionality depends on the electrical and magnetic characteristics of the propagation medium (ϵ_0 and μ_0 for free space).

Another expression that you're likely to encounter involves the average of the power density over time. To understand that, recall from Eq. (5.8) that the time-varying electric field $\vec{E}$ equals $\vec{E}_0 \sin(kz - \omega t)$, so

$$|\vec{E}|^2 = |\vec{E}_0|^2[\sin(kz - \omega t)]^2$$

and the average over time is

$$|\vec{E}|^2_{\text{avg}} = \{|\vec{E}_0|^2[\sin(kz - \omega t)]^2\}_{\text{avg}}.$$

But the average of the $\sin^2$ function over many cycles is 1/2, so

$$|\vec{E}|^2_{\text{avg}} = \frac{1}{2}|\vec{E}_0|^2,$$

which means the average power density is

$$|\vec{S}|_{\text{avg}} = \frac{1}{2}\sqrt{\frac{\epsilon_0}{\mu_0}}|\vec{E}_0|^2. \tag{5.22}$$

The following example illustrates the use of these equations.

Example 5.2 *At the surface of the Earth, the average power density of sunlight on a clear day is approximately 1,300 W/m^2. Find the average magnitude of the electric and magetic fields in sunlight.*

To find the average electric field strength, solve Eq. (5.22) for $|\vec{E}|_{\text{avg}}$:

$$\begin{aligned}
|\vec{S}|_{\text{avg}} &= \frac{1}{2}\sqrt{\frac{\epsilon_0}{\mu_0}}|\vec{E}_0|^2, \\
|\vec{E}_0| &= \sqrt{\frac{2|\vec{S}|_{\text{avg}}}{\sqrt{\epsilon_0/\mu_0}}} = \sqrt{2|\vec{S}|_{\text{avg}}\sqrt{\frac{\mu_0}{\epsilon_0}}} \\
&= \sqrt{(2)1,300\ \text{W/m}^2\sqrt{\frac{4\pi \times 10^{-7}\ \text{H/m}}{8.8541878 \times 10^{-12}\ \text{F/m}}}} \approx 990\ \text{V/m}.
\end{aligned}$$

Once you know the electric-field magnitude, you can use Eq. (5.15) to find the magnitude of the magnetic field:

$$|\vec{B}_0| = \frac{|\vec{E}_0|}{c} = \frac{990\ \text{V/m}}{3 \times 10^8\ \text{m/s}} \approx 3.3 \times 10^{-6}\ \text{T}.$$

The term $\sqrt{\mu/\epsilon}$ is the electromagnetic impedance (usually designed as Z), which plays an analogous role to the mechanical impedance discussed in Chapter 4. In free space, the electromagnetic impedance is

$$Z_0 = \sqrt{\frac{\mu_0}{\epsilon_0}} = \sqrt{\frac{4\pi \times 10^{-7}\ \text{H/m}}{8.8541878 \times 10^{-12}\ \text{F/m}}} \approx 377\,\Omega, \tag{5.23}$$

in which the symbol Ω represents ohms, the SI unit of electromagnetic impedance. The impedance of other materials is given by $Z = \sqrt{\mu/\epsilon}$, where μ is the magnetic permeability of the material and ϵ is the electric permittivity of the material. When an electromagnetic wave impinges on a boundary between materials with different electromagnetic impedance, the amplitudes of the reflected and transmitted waves depend on the impedance difference between the media.

In free space, the power density of an electromagnetic plane wave is related to the magnitude of the wave's electric field by

$$|\vec{S}| = \frac{|\vec{E}|^2}{Z_0} \tag{5.24}$$

and

$$|\vec{S}|_{\text{avg}} = \frac{|\vec{E}|^2_{\text{avg}}}{Z_0}. \tag{5.25}$$

These equations relating the power density of an electromagnetic field to the electric field and the characteristics of the propagation medium are useful, but they're silent as to the direction of the energy flow. As you may have surmised from the notation of power density as the magnitude of the Poynting vector $\vec{S}$, the direction of the Poynting vector tells you the direction of energy flow in an electromagnetic wave.

Here's the most common way of writing the Poynting vector:

$$\vec{S} = \frac{1}{\mu_0}\vec{E} \times \vec{B}, \tag{5.26}$$

in which $\vec{E}$ and $\vec{B}$ are the vector electric and magnetic fields, and the "×" symbol represents the vector cross-product. There are several ways to multiply vectors, but the vector cross-product is exactly what's needed for the Poynting vector. That's because the result of taking a cross-product between two vectors gives a result that's also a vector, and the direction of that vector is perpendicular to the directions of the two vectors in the cross-product.

The magnitude and direction of the vector cross-produce are illustrated in Fig. 5.8. As shown in the figure, for vectors $\vec{A}$ and $\vec{B}$, the magnitude of the cross-product is

$$|\vec{A} \times \vec{B}| = |\vec{A}||\vec{B}|\sin\theta, \tag{5.27}$$

where θ is the angle between $\vec{A}$ and $\vec{B}$.

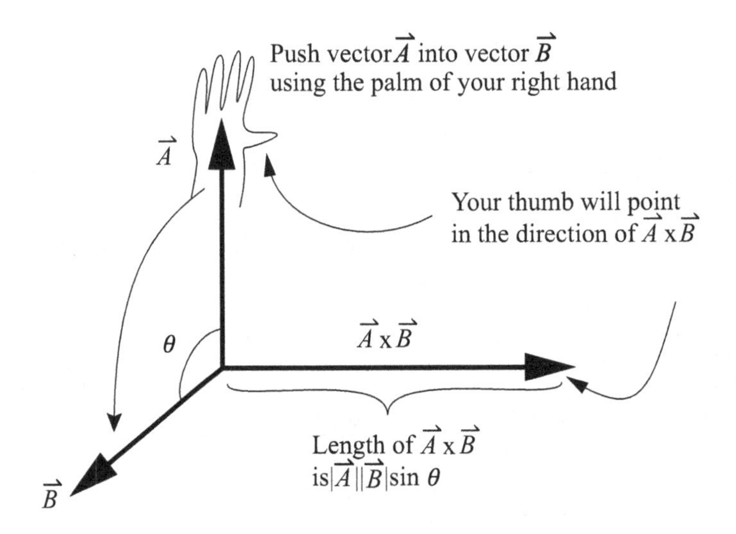

Figure 5.8 The direction of a vector cross-product.

You can determine the direction of the vector cross-product $\vec{A} \times \vec{B}$ using the right-hand rule. To do that, imagine using the palm of your right hand to push the first vector in the cross-product ($\vec{A}$ in this case) into the second vector ($\vec{B}$ in this case), as shown in Fig. 5.8. As you push, keep your right thumb perpendicular to the fingers of your hand. Your right thumb then points in the direction of the cross-product.

Notice that the vector cross-product $\vec{A} \times \vec{B}$ is not the same as $\vec{B} \times \vec{A}$, because the result of pushing vector $\vec{B}$ into vector $\vec{A}$ points in the opposite direction from that shown in Fig. 5.8 (although the magnitude is the same). Thus $\vec{B} \times \vec{A} = -\vec{A} \times \vec{B}$.

If you apply this process to the electric and magnetic fields ($\vec{E}$ and $\vec{B}$) of the electromagnetic wave shown in Fig. 5.5, you'll see that the vector cross-product $\vec{E} \times \vec{B}$ points in the direction of wave propagation (along the z-axis for the wave in Fig. 5.5). Since the propagation direction is perpendicular both to the electric field and to the magnetic field of an electromagnetic plane wave, the cross-product operation provides the proper direction for the Poynting vector.

Example 5.3 *Use the definition of the Poynting vector from Eq. (5.26) to find the vector power density* $\vec{S}$ *of an electromagnetic plane wave propagating along the positive* z*-axis.*

From Fig. 5.5 and Eq. (5.26), the Poynting vector is

$$\begin{aligned}\vec{S} &= \frac{1}{\mu_0}\vec{E} \times \vec{B} \\ &= \frac{1}{\mu_0}|\vec{E}||\vec{B}|\sin\theta\ \hat{k}.\end{aligned}$$

Since $\vec{E}$ is perpendicular to $\vec{B}$, θ is 90°. You also know that $|\vec{B}| = |\vec{E}|/c$, so this can be written as

$$\begin{aligned}\vec{S} &= \frac{1}{\mu_0}|\vec{E}||\vec{B}|\sin 90^\circ\ \hat{k} = \frac{1}{\mu_0}|\vec{E}||\vec{B}|\hat{k} \\ &= \frac{1}{\mu_0}|E|\frac{|\vec{E}|}{c}\hat{k} = \frac{1}{\mu_0}|\vec{E}|\frac{|\vec{E}|}{\sqrt{1/(\mu_0\epsilon_0)}}\hat{k} \\ &= \frac{\sqrt{\mu_0\epsilon_0}}{\mu_0}|\vec{E}|^2\hat{k} = \sqrt{\frac{\epsilon_0}{\mu_0}}|\vec{E}|^2|\hat{k} = \frac{|\vec{E}|^2}{Z_0}\hat{k},\end{aligned}$$

as expected from Eq. (5.24).

5.6 Problems

5.1. If the electric field in a certain region is given by $\vec{E} = 3x^2y\hat{\imath} - 2xyz^2\hat{\jmath} + x^3y^2z^2\hat{k}$ in SI units, what is the electric charge density at the point $x = 2,\ y = 3,\ z = 1$?

5.2. If the magnetostatic field in a certain region is given by $\vec{B} = 3x^2y^2z^2\hat{\imath} + xy^3z^2\hat{\jmath} - 3xy^2z^3\hat{k}$ in SI units, what is the magnitude of the electric current density at the point $x = 1,\ y = 4,\ z = 2$?

5.3. If the magnetic field at a certain location is changing with time according to the equation $\vec{B} = 3t^2\hat{\imath} + t\hat{\jmath}$ in SI units, what are the magnitude and direction of the curl of the induced electric field at that location at time $t = 2$ seconds?

5.4. Show that the x-component of Faraday's law for a plane wave propagating in the positive z-direction leads to the equation $E_{0y} = -cB_{0x}$.

5.5. Show that the y-component of Faraday's law for a plane wave propagating in the positive z-direction leads to the equation $E_{0x} = cB_{0y}$.

5.6. Show that the two equations (5.13) and (5.14) mean that $\vec{E}$ and $\vec{B}$ are perpendicular to one another.

5.7. Show that $\sqrt{1/(\mu_0\epsilon_0)}$ has units of meters per second.

5.8. According to the inverse-square law, the power density of an electromagnetic wave transmitted by an isotropic source (that is, a source that radiates equally in all directions) is given by the equation

$$|\vec{S}| = \frac{P_{\text{transmitted}}}{4\pi r^2},$$

where $P_{\text{transmitted}}$ is the transmitted power and r is the distance from the source to the receiver. Find the magnitude of the electric and magnetic fields produced by a 1,000-watt radio transmitter at a distance of 20 km.

5.9. If vector $\vec{A} = 8\hat{\imath} + 3\hat{\jmath} + 6\hat{k}$ and vector $\vec{B} = 12\hat{\imath} - 7\hat{\jmath} + 4\hat{k}$, what are the magnitude and direction of the vector cross-product $\vec{A} \times \vec{B}$?

5.10. In certain plasmas (such as the Earth's ionosphere), the dispersion relation of electromagnetic waves is $\omega^2 = c^2k^2 + \omega_{\text{p}}^2$, where c is the speed of light and ω_{p} is the natural "plasma frequency" that depends on the particle concentration. Find the phase velocity (ω/k) and the group velocity ($d\omega/dk$) of electromagnetic waves in such a plasma, and show that their product equals the square of the speed of light.

6
The quantum wave equation

This is the third of three chapters that apply the concepts of Chapters 1, 2, and 3 to the three principal types of waves: mechanical, electromagnetic, and quantum. Although mechanical waves may be the most obvious in everyday life, and electromagnetic waves may be the most useful to our technological society, a case can be made that quantum waves are the most fundamental. As you'll see in this chapter, every bit of matter in the Universe behaves like a wave under certain circumstances, so it's hard to imagine anything more fundamental.

If you're thinking that life in the macroscopic world moves along just fine without considering quantum effects, you should realize that the laws of classical physics are approximations, and all the quantum weirdness described in this chapter blends smoothly with classical physics when they overlap. Much of that weirdness comes from the dual wave and particle nature of matter and energy, and one of the goals of this chapter is to help you understand what's actually doing all this waving.

This chapter begins with a comparison of the characteristics of waves and particles in Section 6.1, followed by a discussion of wave–particle duality in Section 6.2. You can read about the Schrödinger equation and probability wavefunctions in Sections 6.3 and 6.4, and quantum wave packets are discussed in Section 6.5.

6.1 Wave and particle characteristics

Before studying modern physics, most students think of particles and waves as belonging to fundamentally different categories of objects. That's understandable, because particles and waves have several very different characteristics, including the way in which they occupy space, the way they travel through openings, and the way they interact with other particles or waves. Here's a summary of some of the differentiating characteristics between particles and waves.

Occupying space. Particles exist in a well-defined amount of space; if you could pause time at some instant during a tennis match, all of the spectators would agree on the position of the ball at that instant. This property means particles are "localized", since the particle is in a particular location at a given time. Waves, on the other hand, exist over an extended region of space. As described in Section 3.3 of Chapter 3, a harmonic function such as $\sin(kx - \omega t)$ exists over all values of x, from $-\infty$ to $+\infty$, with no way to distinguish one cycle from another. So a single-frequency wave is inherently non-localized; it exists everywhere.

These differences are illustrated in Fig. 6.1, which shows a particle represented by the small circle and a wave represented by the curving line at one instant in time (so this is a "snapshot" of the particle and wave). At the instant of the snapshot, all observers can agree that the particle exists at position

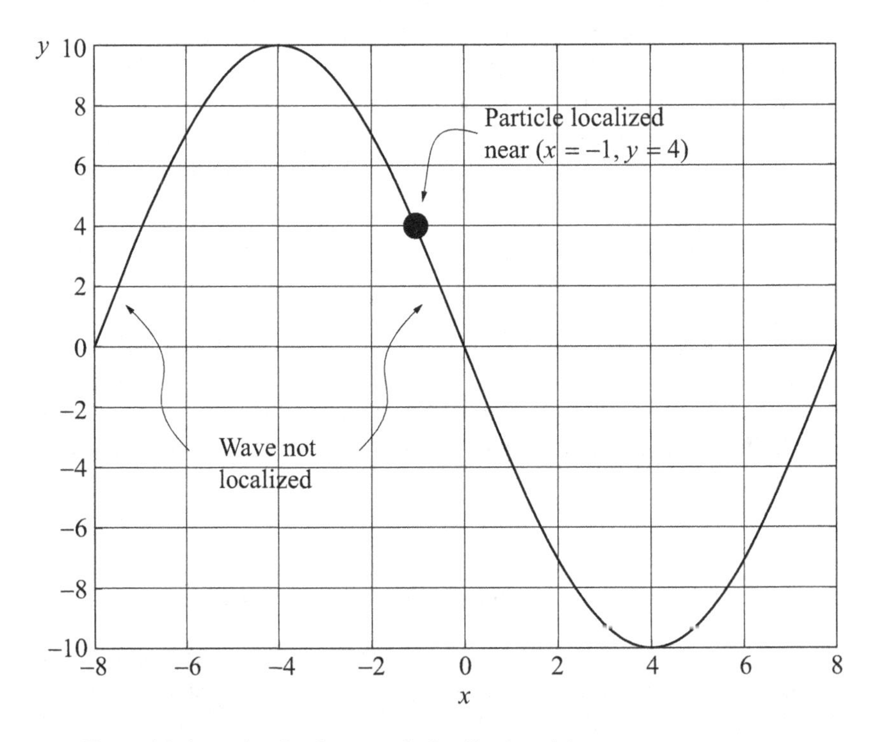

Figure 6.1 A non-localized wave and a localized particle.

$x = -1$ and $y = 4$. But where is the wave at that instant? The wave exists at $x = -8$, $x = 0$, $x = 8$, and all other values of x (including locations not shown to the left and right of the graph and in between the x values listed here). Any given peak, trough, or zero-crossing of the wave exists at a specific location at a certain time, but every cycle looks exactly like every other cycle, and the wave itself consists of all its portions. Hence the wave is non-localized.

Traveling through openings. When a particle passes through an opening larger than the particle, its path is unaffected, as shown in the upper left portion of Fig. 6.2. Only when the size of the opening is smaller than the size of the particle is the particle's behavior affected by the presence of the opening.

Waves, however, behave quite differently. The right portion of Fig. 6.2 shows the effect of openings of various sizes on a plane wave incident from the left. In the top right portion of the figure, the aperture is many times larger than the wavelength of the wave, and the wave passes through the opening without significant change. But, if you reduce the size of the opening so that the size of the aperture is similar to the wavelength of the wave, the portion of the wave

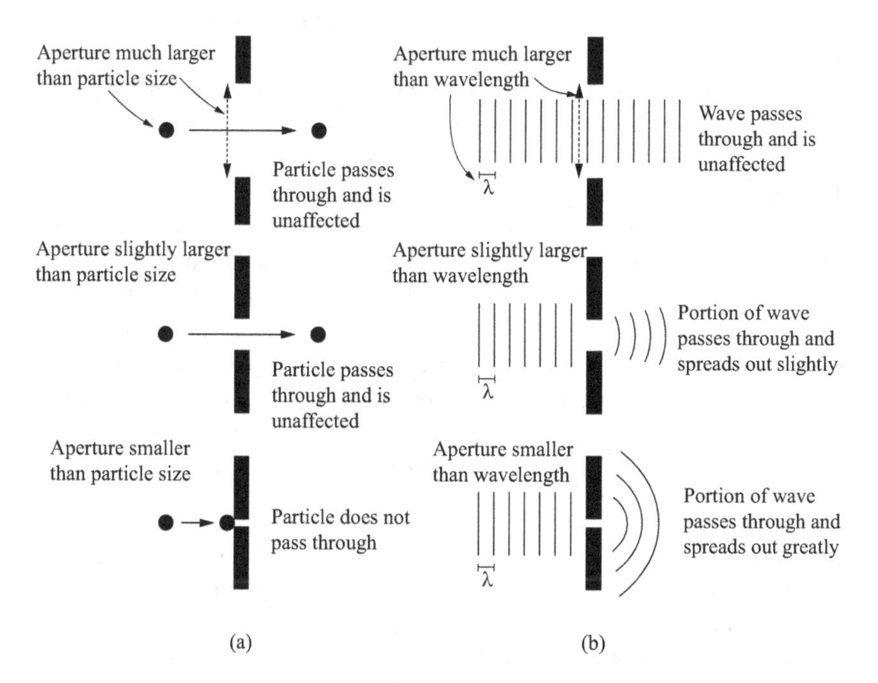

Figure 6.2 (a) Particles are unaffected by an opening as long as they fit through. (b) Waves spread out beyond the opening when the aperture is comparable to the wavelength or smaller.

that passes through the opening is no longer a plane wave. As you can see in the center portion of the figure, the surfaces of constant phase are somewhat curved, and the wave spreads out after passing through the opening. And, if you reduce the size of the opening even more, such that the aperture is smaller than the wavelength of the wave, the wavefront curvature and spreading out of the wave becomes even greater. This effect is called "diffraction".

Interacting with other particles and waves. Particles interact with other particles via collisions, and when particles collide they can exchange momentum and energy. A bowling ball does just that when hitting pins – the system's "before" state is a fast ball and stationary pins, while the "after" state is a slower ball and pins that fly quickly away from their original positions. According to conservation laws, the total momentum and energy of the ball–pins system is the same before and after, but the energy and momentum of each object may change as a result of the collision. Such collisions happen quickly (instantaneously in the ideal case of no deformation), and this quick exchange of a discrete amount of energy and momentum is a hallmark of particle interactions.

Waves, however, interact with other waves via superposition rather than by collisions. As described in Section 3.3 of Chapter 3, when two or more waves occupy the same region of space, the resulting wave is the sum of all contributing waves. The waves are said to "interfere" with one another, and whether the resultant has larger amplitude (constructive interference) or smaller amplitude (destructive interference) than the contributing waves depends on the relative phases of those waves.

Waves can also interact with objects, and in such interactions some of the wave's energy can be transferred to the object. For example, imagine a buoy bobbing up and down on the surface of a lake as a wave passes by. When the wave lifts the buoy, both the kinetic and the gravitational potential energy of the buoy increase, and the energy acquired by the buoy comes directly from the wave. Unlike the near-instantaneous energy transfer that occurs when particles collide, the transfer of energy from a wave to an object occurs over an extended period of time. The rate of energy transfer depends on the characteristics of the wave.

A summary of the classical-mechanics model of some of the distinctions between particles and waves is given in the diagram in Fig. 6.3. In the early part of the twentieth century, a series of novel experiments and theoretical advances began to suggest that new mechanics were needed, and those "quantum mechanics" incorporate a very different understanding of particles and waves. In modern physics, particles can exhibit wave behavior and waves can act like particles. This "wave–particle duality" is the subject of the next section.

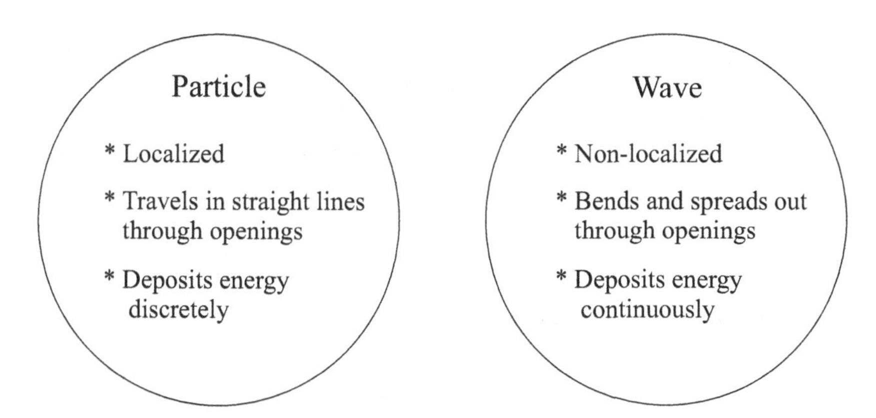

Figure 6.3 A model of classical particle and wave differences.

6.2 Wave–particle duality

To understand wave–particle duality, it's helpful to consider the type of experiment you could perform on a particle to test for wave behavior. As discussed in the previous section, one classical difference between particles and waves is how they pass through an opening such as a slit in an opaque barrier. If a stream of particles (or even a single particle) passing through a slit exhibits diffraction, you know that those particles are behaving like waves. However, in order for the diffraction to be significant, the size of the aperture must be comparable to the wavelength of the wave. So, if you hope to use diffraction to detect a particle's wave behavior, you must construct an aperture that matches the particle's wavelength. That means that it's necessary to predict the wavelength from the properties of the particle.

That prediction is possible due to the work of French physicist Louis de Broglie, who hypothesized the existence of matter waves in 1924. Almost two decades before, Albert Einstein had shown that light exhibits characteristics of both waves and particles, behaving like waves when passing through slits, but acting like particles (called photons) during interactions such as the photoelectric effect. De Broglie made the inspired guess that this wave–particle duality extends to electrons and other particles with mass. As he wrote in his PhD thesis: "My essential idea was to extend to all particles the coexistence of waves and particles discovered by Einstein in 1905 in the case of light and photons."

To understand de Broglie's expression for the wavelength of a particle, consider Einstein's relationship for the energy of a photon with frequency ν and wavelength λ:

$$E = h\nu = \frac{hc}{\lambda}, \tag{6.1}$$

where h is Planck's constant (6.626×10^{-34} m^2 kg/s) and c is the speed of light. Although photons have no rest mass, they do carry momentum, and the magnitude of that momentum is related to the photon's energy by

$$p = \frac{E}{c}. \tag{6.2}$$

Thus, for a photon, $E = cp$, and substituting this into Eq. (6.1) gives

$$E = \frac{hc}{\lambda} = cp,$$
$$p = \frac{h}{\lambda},$$
$$\lambda = \frac{h}{p}.$$

If this relationship holds also for particles with mass, then the wavelength of a particle of mass m traveling at speed v is

$$\lambda = \frac{h}{p} = \frac{h}{mv}, \tag{6.3}$$

since the magnitude of the particle's momentum is $p = mv$.

Consider the implications of Eq. (6.3). Because wavelength and momentum are inversely proportional, the wavelength of a particle increases as the particle's momentum decreases. This is why only moving particles exhibit wave behavior – at zero velocity, the wavelength becomes infinite and unmeasurable. Also, because Planck's constant is such a small number, wave behavior is measurable only when the momentum is small enough to make the total fraction h/p reasonably large. Thus the wave characteristics of everyday objects are not apparent because their high mass makes their de Broglie wavelengths too small to measure.

Here are two examples that will give you an idea of the size of the de Broglie wavelengths for objects of various masses.

Example 6.1 *What is the de Broglie wavelength of a 75-kg human walking at a speed of 1.5 m/s?*

The human in this example has momentum $p = 75$ kg $\times$ 1.5 m/s $= 113$ kg m/s, which gives a de Broglie wavelength of

$$\lambda = \frac{6.626 \times 10^{-34}\ \text{Js}}{113\ \text{kg m/s}} = 5.9 \times 10^{-36}\ \text{m}. \tag{6.4}$$

This is not only billions of times smaller than the spacing between atoms in a typical solid, but billions of times smaller than the protons and neutrons that make up the atoms' nuclei. Hence an object with the mass of a human is not a good candidate for demonstrating the wave behavior of matter. However, for a very small mass with very low velocity, the de Broglie wavelength can be large enough to be measured.

Example 6.2 *What is the de Broglie wavelength of an electron that has passed through a potential difference of 50 volts?*

After passing through a potential difference of 50 volts, an electron has 50 electron volts (eV) of energy. One eV is 1.6×10^{-19} J, so the electron's energy in SI units is

$$50\text{ eV}\,\frac{1.6\times 10^{-19}\text{ J}}{1\text{ eV}} = 8\times 10^{-18}\text{ J}. \tag{6.5}$$

To relate this energy to the electron's momentum, you can use the classical expression for kinetic energy,

$$\text{KE} = \frac{1}{2}mv^2, \tag{6.6}$$

and then multiply and divide the right-hand side by the mass:

$$\text{KE} = \frac{1}{2}mv^2 = \frac{1}{2m}m^2v^2 = \frac{p^2}{2m}. \tag{6.7}$$

In this case, the electron's energy (E) is all kinetic, so $E = \text{KE}$, and the momentum is

$$p = \sqrt{2mE}. \tag{6.8}$$

Applying this to the electron with 8×10^{-18} J of energy, the momentum is

$$\begin{aligned} p &= \sqrt{2\times 9.1\times 10^{-31}\text{ kg}\times 8\times 10^{-18}\text{ J}} \\ &= 3.8\times 10^{-24}\text{ kg m/s}. \end{aligned}$$

Putting this result into de Broglie's equation (Eq. (6.3)) gives a wavelength of

$$\lambda = \frac{6.626\times 10^{-34}\text{ J s}}{3.8\times 10^{-24}\text{ kg m/s}} = 1.7\times 10^{-10}\text{ m}, \tag{6.9}$$

or 0.17 nanometers. This is similar to the spacing between atoms in a crystal array, so such an array can be used to experimentally determine a moving electron's wavelength.

That experiment was done by Clinton Davisson and Lester Germer in 1927. Knowing that waves scattered from crystals produce a diffraction pattern, Davisson and Germer bombarded a nickel crystal with electrons and looked for evidence of diffraction. They found it, and then they used the scattering angle and known spacing of the atoms in the crystal to calculate the wavelength of the electrons. Their results were in good agreement with de Broglie's expression for the wavelength of particles with the mass and velocity of the electrons used in the experiment.

A similar, but conceptually more straightforward, experiment is the double slit, which is often used to demonstrate the wave properties of light. You can find the details of this important experiment in any comprehensive optics text, but the following overview should help you understand how this experiment can be used to demonstrate the wave nature of particles.

The double slit is set up as in Fig. 6.4 with a source of waves or particles facing a barrier with two small slits a short distance apart; far behind the barrier is a detector.

First consider what happens when the source emits a continuous stream of waves in the direction of the barrier. One portion of the wavefront arrives at the left slit and another portion arrives at the right slit. According to Huygens' principle, every point on a wavefront can be considered to be the source of another wave that spreads out spherically from that point. If no barrier were present, all of the spherical waves from the Huygens sources on one planar wavefront

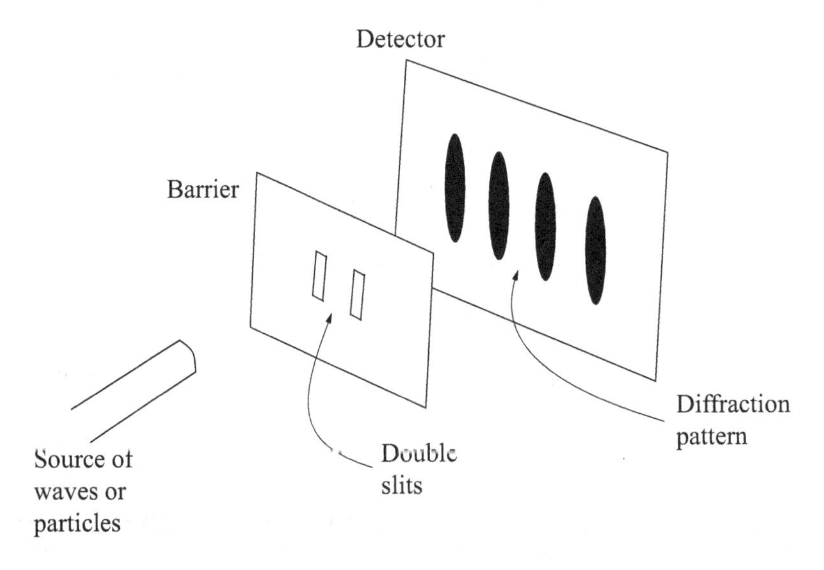

Figure 6.4 The double-slit experimental set-up.

would superimpose to produce the next wavefront. All the points on that wavefront can also be considered to be Huygens sources, emitting another set of spherical waves that will add up to produce the next downstream wavefront.

But, when the wave encounters the barrier, most of the secondary waves are blocked; only those that originate on two small portions of the wavefront pass through the slits. The spherically expanding secondary waves from those two portions of the wavefront then travel to the detector, where they superimpose to produce the resultant waveform. But since the distance from the left slit to the detector is not the same as the distance from the right slit to the detector, the waves from the two slits arrive at the detector with different phases. Depending on the amount of that phase difference, the two waves may add or subtract, and the interference may be constructive or destructive. At points of constructive interference, a bright fringe appears; and at points of destructive interference, a dark fringe appears. Such bright and dark fringes are the hallmark of wave behavior.

Now imagine that the source fires a stream of particles such as electrons at the barrier. Some of those electrons will strike the barrier, and others will pass through the slits. If electrons were simply particles with no wave characteristics, the detector would register accumulations of discrete energy deposited roughly in the shape of the slits, as shown in the left portion of Fig. 6.5. But, since streaming electrons also have characteristics of waves, the detector registers a continuous accumulation of energy in an interference pattern, as shown in the right portion of Fig. 6.5.

Finally, consider what happens when individual electrons (rather than a stream) are sent through the slits one at a time. In that case, the detector

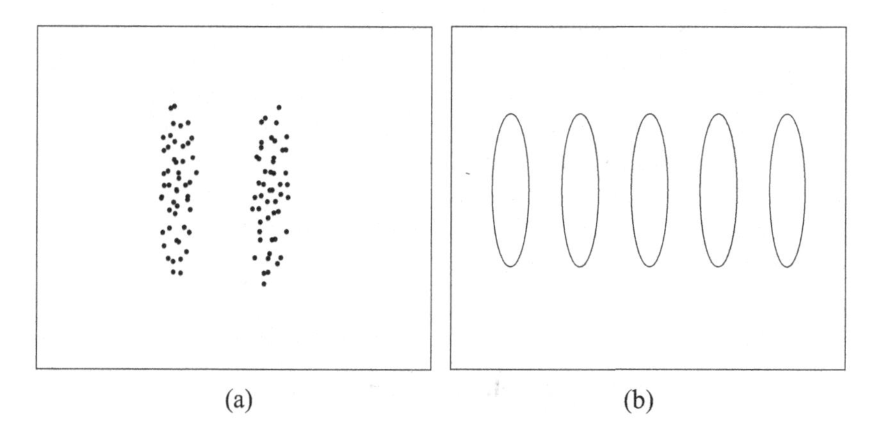

Figure 6.5 (a) Classical particle results for the double slit and (b) wave results for the double slit.

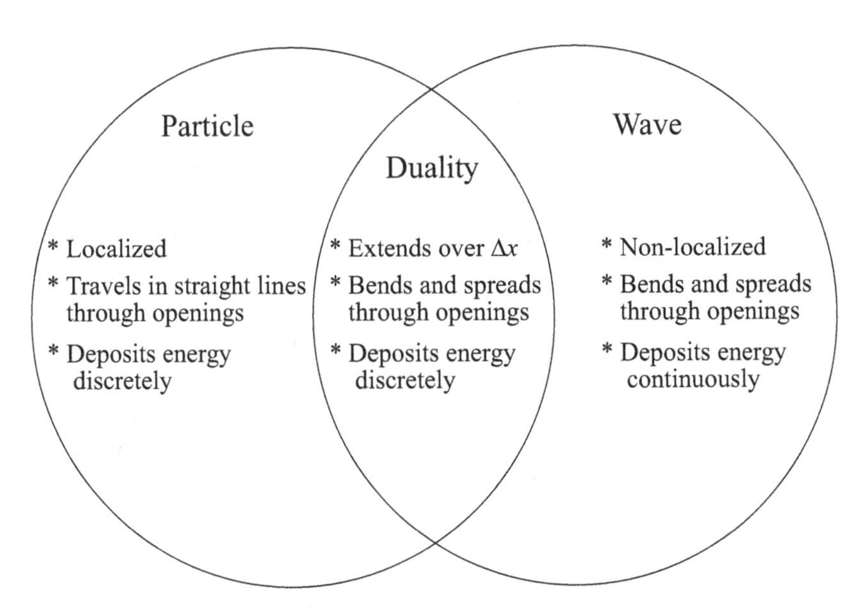

Figure 6.6 A model of quantum particle and wave differences and the wave–particle duality.

registers each electron as a single point, but the accumulation of points creates the dark and light fringes of an interference pattern. Evidently each electron interferes with itself as it passes through *both* slits.

The modern understanding of wave–particle duality is shown in the diagram in Fig. 6.6. When a quantum-mechanical object is traveling, it acts like a wave. This includes bending around corners, diffracting through small openings, and interfering. When a quantum-mechanical object is depositing or accepting energy, it acts like a particle. Instead of being completely localized, existing at a single point like a particle, or completely non-localized, being spread out over all space, a quantum-mechanical object behaves like a traveling wave packet that occupies a small, but finite, amount of space. To understand why quantum objects behave this way – as opposed to, say, traveling like a particle and interacting like a wave – it helps to understand the Schrödinger equation, which is the subject of the next section.

6.3 The Schrödinger equation

Understanding that electrons can behave like waves, you may be wondering "Exactly what is waving?" Or, to ask it another way, why do electrons travel like a wave and interact like a particle? The answer comes from the interpretation of the wavefunction, $\Psi(x, t)$, the solution to the Schrödinger equation which governs these matter waves.

Although the great physicist Richard Feynman once said that it's not possible to derive the Schrödinger equation from anything you know, you can get a sense of Schrödinger's reasoning by starting with an expression for the energy of a particle. If that particle is moving non-relativistically (that is, slowly relative to the speed of the light), and the forces acting on the particle are conservative (so that a potential energy may be associated with each force), the total mechanical energy (E) of the particle may be written as

$$E = \text{KE} + V = \frac{1}{2}mv^2 + V = \frac{p^2}{2m} + V, \tag{6.10}$$

where KE is the kinetic energy and V is the potential energy of the particle.

Since photons (which can be considered to be "particles of light") of frequency ν have energy $E = h\nu$ (Eq. (6.1)), Schrödinger wrote a version of this equation for the energy associated with matter waves. You'll see that below, but first you should make sure you're aware of the difference between the Planck constant h and the "reduced Planck constant" $\hbar$ as well as the relationship of the frequency ν to the angular frequency ω:

$$E = h\nu = \left(\frac{h}{2\pi}\right)(2\pi\nu) = \hbar\omega, \tag{6.11}$$

where $\hbar = h/(2\pi)$ and $\omega = 2\pi\nu$. Inserting this expression for energy into the left-hand side of Eq. (6.10) and using $p = mv$ gives

$$\hbar\omega = \frac{p^2}{2m} + V. \tag{6.12}$$

De Broglie's equation relating momentum to wavelength is

$$p = \frac{h}{\lambda} = \frac{h}{2\pi/k} = \left(\frac{h}{2\pi}\right)k = \hbar k, \tag{6.13}$$

so you can write Eq. (6.12) as

$$\hbar\omega = \frac{\hbar^2 k^2}{2m} + V. \tag{6.14}$$

To get from this form of the energy equation to the Schrödinger equation, assume that the matter wave associated with the particle can be written as the harmonic wavefunction $\Psi(x, t)$:

$$\Psi(x, t) = Ae^{i(kx-\omega t)} = e^{-i\omega t}(Ae^{ikx}). \tag{6.15}$$

Notice that we've divided the exponential to separate the time-dependent term from the space-dependent term. That makes it a bit easier to take some derivatives that show how this wavefunction fits into Eq. (6.14).

The first of those derivatives is the derivative of Ψ with respect to time (t):

$$\frac{\partial \Psi}{\partial t} = \frac{\partial (e^{-i\omega t})}{\partial t}(Ae^{ikx}) = -i\omega(e^{-i\omega t})(Ae^{ikx}) = -i\omega\Psi. \tag{6.16}$$

Now consider the derivative of Ψ with respect to x,

$$\frac{\partial \Psi}{\partial x} = e^{-i\omega t}\frac{\partial (Ae^{ikx})}{\partial x} = ike^{-i\omega t}(Ae^{ikx}) = ik\Psi, \tag{6.17}$$

and the second derivative with respect to x,

$$\frac{\partial^2 \Psi}{\partial x^2} = e^{-i\omega t}\frac{\partial^2 (Ae^{ikx})}{\partial x^2} = (ik)^2 e^{-i\omega t}(Ae^{ikx}) = -k^2\Psi. \tag{6.18}$$

The last step before fitting these derivatives into Eq. (6.14) is to multiply Eq. (6.16) by $i\hbar$:

$$i\hbar\frac{\partial \Psi}{\partial t} = (i\hbar)(-i\omega\Psi) = \hbar\omega\Psi$$

or

$$\hbar\omega = \frac{i\hbar}{\Psi}\left(\frac{\partial \Psi}{\partial t}\right).$$

Now you can plug this into Eq. (6.14):

$$\frac{i\hbar}{\Psi}\left(\frac{\partial \Psi}{\partial t}\right) = \frac{\hbar^2 k^2}{2m} + V$$

or

$$i\hbar\left(\frac{\partial \Psi}{\partial t}\right) = \frac{\hbar^2 k^2}{2m}\Psi + V\Psi.$$

But you know that $k^2\Psi = -\partial^2\Psi/\partial x^2$ from Eq. (6.18), so

$$i\hbar\left(\frac{\partial \Psi}{\partial t}\right) = \frac{-\hbar^2}{2m}\frac{\partial^2 \Psi}{\partial x^2} + V\Psi. \tag{6.19}$$

This is the one-dimensional time-dependent Schrödinger equation. As mentioned in Section 2.4 of Chapter 2, the Schrödinger equation differs from the classical wave equation in that the partial derivative with respect to time is a first rather than a second derivative, which has important implications for the nature of the solutions. Note also that the presence of "i" as a multiplicative factor means that the solutions will generally be complex.

Before considering those solutions, you should be aware that you're likely to encounter a "time-independent" version of the Schrödinger equation. You can derive that version from the time-dependent version by using Eq. (6.16) to replace the time derivative $\partial\Psi/\partial t$ with $-i\omega\Psi$:

$$i\hbar(-i\omega\Psi) = \frac{-\hbar^2}{2m}\frac{\partial^2\Psi}{\partial x^2} + V\Psi$$

or

$$\hbar\omega\Psi = \frac{-\hbar^2}{2m}\frac{\partial^2\Psi}{\partial x^2} + V\Psi. \tag{6.20}$$

Since $E = \hbar\omega$, you may also see this equation written as

$$E\Psi = \frac{-\hbar^2}{2m}\frac{\partial^2\Psi}{\partial x^2} + V\Psi \tag{6.21}$$

or

$$(E - V)\Psi = \frac{-\hbar^2}{2m}\frac{\partial^2\Psi}{\partial x^2}. \tag{6.22}$$

It's important for you to note that the label "time-independent" does not mean that the wavefunction Ψ is not a function of time; Ψ is still $\Psi(x, t)$. So what's independent of time in this case? It's the energy terms on the left-hand side of Eq. (6.22).

Before diving into the solutions to the Schrödinger equation, think about the meaning of Eq. (6.22). It's essentially an expression of the conservation of energy: The total energy minus the potential energy equals the kinetic energy. So Eq. (6.22) tells you that the kinetic energy of the particle is proportional to the the second spatial derivative of Ψ, which is the curvature of the wavefunction. Greater curvature with x means that the waveform has a higher spatial frequency (it goes from positive to negative in a shorter distance), which means it has shorter wavelength, and de Broglie's equation relates shorter wavelength to higher momentum.

Example 6.3 *What is the time-independent Schrödinger equation for a free particle?*

In this context, "free" means that the particle is free of the influence of external forces, and, since force is the gradient of potential energy, a free particle travels in a region of constant potential energy. Since the reference location of zero potential energy is arbitrary, you can set $V = 0$ in the Schrödinger equation for a free particle. Thus Eq. (6.22) becomes

$$E\Psi = \frac{-\hbar^2}{2m}\frac{\partial^2\Psi}{\partial x^2} \tag{6.23}$$

or

$$\frac{\partial^2\Psi}{\partial x^2} = -\frac{2mE}{\hbar^2}\Psi. \tag{6.24}$$

Since the total energy of a free particle equals the particle's kinetic energy, you can set $E = p^2/(2m)$:

$$\frac{\partial^2\Psi}{\partial x^2} = -\frac{p^2}{\hbar^2}\Psi. \tag{6.25}$$

This is the Schrödinger equation for a free particle. It's instructive to compare this equation with the equation for a standing wave (Eq. (3.22) in Section 3.2 of Chapter 3). That equation is

$$\frac{\partial^2 X}{\partial x^2} = \alpha X = -k^2 X, \tag{3.22}$$

which is identical to Eq. (6.25), provided that

$$\frac{p^2}{\hbar^2} = k^2.$$

But the wavenumber $k = 2\pi/\lambda$, so

$$\frac{p^2}{\hbar^2} = \left(\frac{2\pi}{\lambda}\right)^2$$

or

$$\frac{p}{\hbar} = \frac{p}{h/(2\pi)} = \left(\frac{2\pi}{\lambda}\right).$$

Solving this equation for λ gives

$$\lambda = \frac{h}{p},$$

which is de Broglie's expression for the wavelength of a matter wave. For a free particle, the oscillation frequency and energy may take on any value, but particles in regions in which the potential energy varies may be constrained to certain values of frequency and energy that depend on the boundary conditions. Thus the allowed energies of a particle in a potential well are quantized, as you can see in the chapter-end problems and online solutions.

This means that the wave characteristic of a particle is analogous to a standing wave, and the energy of the particle is proportional to the oscillation frequency of that standing wave. You may see solutions to the time-independent Schrödinger equation referred to as "stationary states", but you should note that this means only that the energy of the system is constant over time, not that the wavefunction is stationary.

6.4 Probability wavefunctions

When he first wrote down his equation, Schrödinger didn't know what the wavefunction $\Psi(x, t)$ physically represented; he guessed that it was the charge density of the electron. This is a reasonable guess, since he was writing down an equation for electrons to follow, and, as groups of electrons are spread through space, you can evaluate the amount of charge per unit volume (the charge density) as a way of knowing where electrons are congregated or sparse. However, Max Born was able to show that this view is inconsistent with experiments and offered his own explanation, which is now the modern understanding.[1]

The modern interpretation of the wavefunction is that it is a "probability amplitude" related to the probability of finding the particle in a given region of space. This quantity is called an amplitude because, just as you must square the amplitude of a mechanical wave to get its energy, so you must square the wavefunction to get a probability density ($\mathcal{P}$). Since the wavefunction is generally complex, you square it by multiplying it by its complex conjugate:

$$\mathcal{P}(x, t) = \Psi^*(x, t)\Psi(x, t) \tag{6.26}$$

or

$$\mathcal{P}(x, t) = |\Psi(x, t)|^2. \tag{6.27}$$

In one dimension the probability density is the probability per unit length, in two dimensions it's the probability per unit area, and in three dimensions it's the probability per unit volume. In other words, $\mathcal{P}(x, t)$ tells you how the probability of finding the particle in a particular place is spread out through space at any given time.

This answers the question of what's waving for an electron and other quantum-mechanical objects: A traveling particle is actually a traveling lump of probability amplitude. When it encounters an obstacle (such as the two slits described in Section 6.2), the waving probability amplitude is diffracted according to the wavelength. When it interacts (which is how it is measured or

[1] In fact, the correspondence between the wavefunction and the charge density is given by the Kohn–Sham equations and is the foundation of density-functional theory, so Schrödinger wasn't entirely off the mark.

detected), this wavefunction collapses to the single measured result. This result is discrete, which is consistent with particle behavior (such as the individual dots that appear on the detector in the electron version of the double-slit experiment).

So how can we write the wavefunction for a free particle? Earlier we used an initial guess of the complex harmonic function,

$$\Psi(x, t) = Ae^{i(kx-\omega t)}, \tag{6.28}$$

which is the complex combination of a cosine and sine wave, as shown in Section 1.4. But a problem arises when you try to calculate the probability of finding the particle somewhere in space using this waveform. Since $\mathcal{P}(x, t)$ represents the probability density, integrating over all space ($x = -\infty$ to $x = +\infty$) should give a probability of one (since the chance of finding the particle somewhere in space is 100%). However, using the wavefunction of Eq. (6.28), that integration looks like this:

$$1 = \int_{-\infty}^{\infty} \Psi^*(x, t)\Psi(x, t)dx \tag{6.29}$$

$$1 = \int_{-\infty}^{\infty} A^*Ae^{-i(kx-\omega t)}e^{i(kx-\omega t)}dx \tag{6.30}$$

$$1 = A^*A(\infty). \tag{6.31}$$

Since nothing can multiply infinity to give one, this waveform is "non-normalizable" (in this context, "normalization" refers to the process of scaling the wavefunction to set the probability of finding the particle somewhere in space to 100%). To modify the wavefunction of Eq. (6.28) to make it normalizable, the Fourier concepts introduced in Section 3.3 of Chapter 3 are very helpful. You can see how to use those concepts to form quantum wave packets in the next section.

6.5 Quantum wave packets

A particle is localized in space, so it seems reasonable to expect that its wave should also be spatially limited; that is, it should be a wave packet rather than a single-wavelength wave with constant amplitude over all space. Ideally this wave packet should be dominated by a particular wavelength (or momentum) so that the de Broglie hypothesis is still roughly applicable. But, as described in Section 3.3 of Chapter 3, it's impossible to form a wave packet without including some amount of waves with different (but similar) wavelengths. A range of wavelengths means a range of wavenumbers (since $k = 2\pi/\lambda$), and a range of wavenumbers means a range of momenta (since $p = \hbar k$). So the challenge is to make a wave packet that is localized in space over a region Δx but that travels with a well-defined momentum $p = \hbar k_0$, where k_0 represents the dominant wavenumber.

Such a wavefunction $\Psi(x, t)$ depends on both location (x) and time (t), but it's a bit easier to see what's going on by separating the variables as described in Section 3.2. Writing $\Psi(x, t) = f(t)\psi(x)$ allows you to concentrate on localizing the spatial term $\psi(x)$; the effect of the time term $f(t)$ will be considered later in this section.

One approach to limiting the spatial extent of the wavefunction is to write $\psi(x)$ as the product of two functions: the "exterior" envelope $g(x)$, and the "interior" oscillations $f(x)$,

$$\psi(x) = g(x)f(x). \tag{6.32}$$

If the envelope function goes to zero everywhere except at a certain range of x values, then the oscillations of the wave packet will be localized to that range.

For example, consider the single-wavelength oscillating function $f(x) = e^{ikx}$. As you can see the plot of the real part of $f(x)$ in Fig. 6.7(a) (for which we've arbitrarily picked $k = 10$), this function extends over all space (from $x = -\infty$ to $x = +\infty$).

Now consider the envelope function $g(x)$ given by the equation

$$g(x) = e^{-ax^2}. \tag{6.33}$$

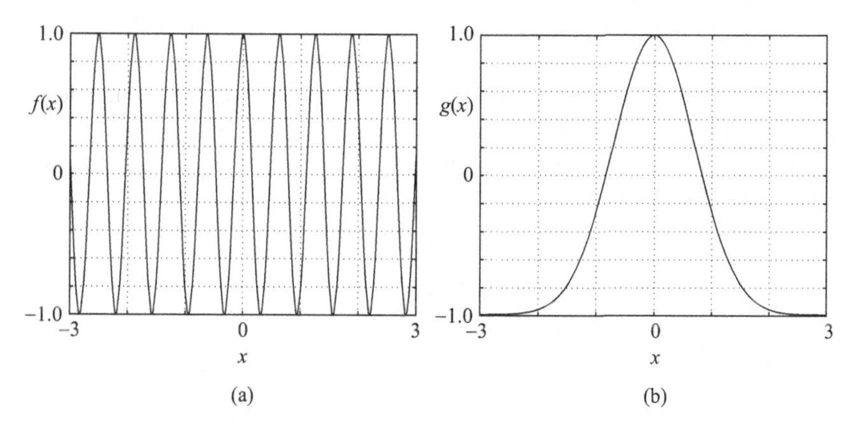

Figure 6.7 (a) The real part of the oscillating function $f(x) = e^{ikx}$ and (b) the envelope function $g(x) = e^{-ax^2}$.

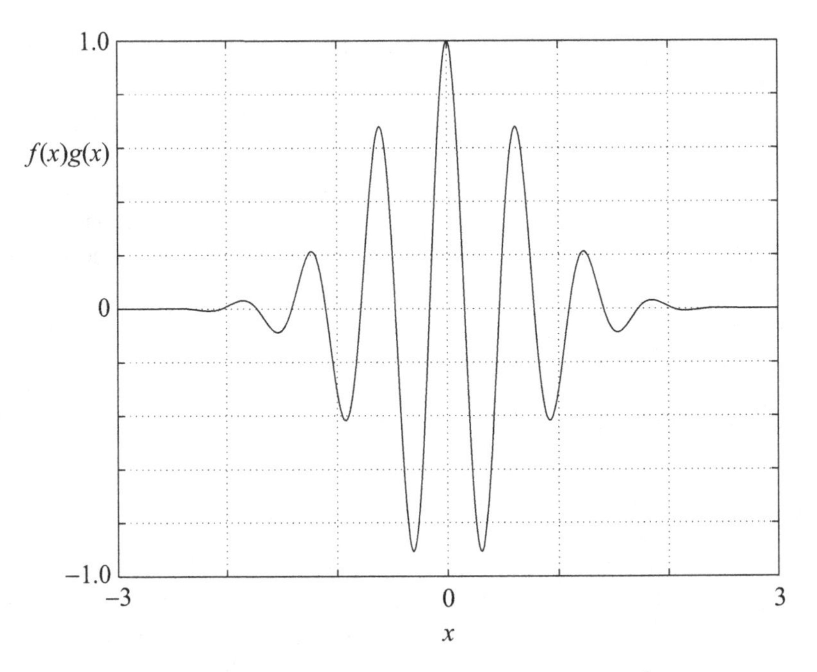

Figure 6.8 The real part of the product function $f(x)g(x) = e^{ikx}e^{-ax^2}$.

As you can see in Fig. 6.7(b), this function reaches a peak of $g(x) = 1$ at $x = 0$ and decreases toward zero in both directions at a rate determined by the value of the constant a (we've selected $a = 1$ for this figure).

Multiplying the envelope function $g(x)$ by the oscillating function $f(x)$ causes the product function $f(x)g(x)$ to roll off both in the negative direction and in the positive direction, as shown in the plot of the real part of $f(x)g(x)$ in Fig. 6.8. This function no longer has a single wavelength (if it did, it wouldn't roll off over distance); you can read more about its wavelength content below. But first you should consider the probability density of $f(x)g(x)$.

Since $f(x) = e^{ikx}$ and $g(x) = e^{-ax^2}$, the wavefunction ψ and its complex conjugate ψ^* are

$$\psi = e^{ikx}e^{-ax^2},$$
$$\psi^* = e^{-ikx}e^{-ax^2},$$

so the probability density is

$$\mathcal{P} = |\psi^*\psi| = \left(e^{-ikx}e^{-ax^2}\right)\left(e^{ikx}e^{-ax^2}\right)$$
$$= e^{-2ax^2}.$$

Integrating this over all space gives

$$\mathcal{P}_{\text{all space}} = \int_{-\infty}^{\infty} e^{-2ax^2}\,dx$$
$$= \sqrt{\frac{\pi}{2a}}.$$

So to set $\mathcal{P}_{\text{all space}} = 1$, you're going to have to scale $\psi^*\psi$ by the inverse of this factor. That means that the function ψ must be scaled by the square root of the inverse of $\sqrt{\pi/(2a)}$, so

$$\psi(x) = \sqrt{\frac{1}{\sqrt{\pi/(2a)}}}e^{-ax^2}e^{ikx} = \left(\frac{2a}{\pi}\right)^{1/4} e^{-ax^2}e^{ikx}. \tag{6.34}$$

This function has the desired characteristics of limited spatial extent while oscillating at a dominant wavelength, and it's normalized to give an all-space probability of one.

You can see how this works for a specific value of the width constant a in the following example.

Example 6.4 *Determine the probability of finding a particle at a given location if the particle's wavefunction is defined as*

$$\psi(x) = \left(\frac{0.2}{\pi}\right)^{1/4} e^{-0.1x^2}e^{ikx}.$$

In this case, the width constant a is 0.1, which makes the probability density

$$\psi^*(x)\psi(x) = \left[\left(\frac{0.2}{\pi}\right)^{1/4} e^{-0.1x^2}e^{-ikx}\right]\left[\left(\frac{0.2}{\pi}\right)^{1/4} e^{-0.1x^2}e^{ikx}\right]$$
$$= \left(\frac{0.2}{\pi}\right)^{1/2} e^{-0.2x^2},$$

which is a Gaussian distribution, as shown in Fig. 6.9.

To find the probability of the particle with this wavefunction being located in a particular spot, you have to integrate the density around that place. In this example, the likelihood of finding the particle at $x = 1$ m, give or take 0.1 m, is

$$\mathcal{P}(1 \pm 0.1) = \left(\frac{0.2}{\pi}\right)^{1/2} \int_{0.9}^{1.1} e^{-0.2x^2}\,dx$$
$$= 0.041,$$

or 4.1%. You can check the normalization of this function by integrating over all space:

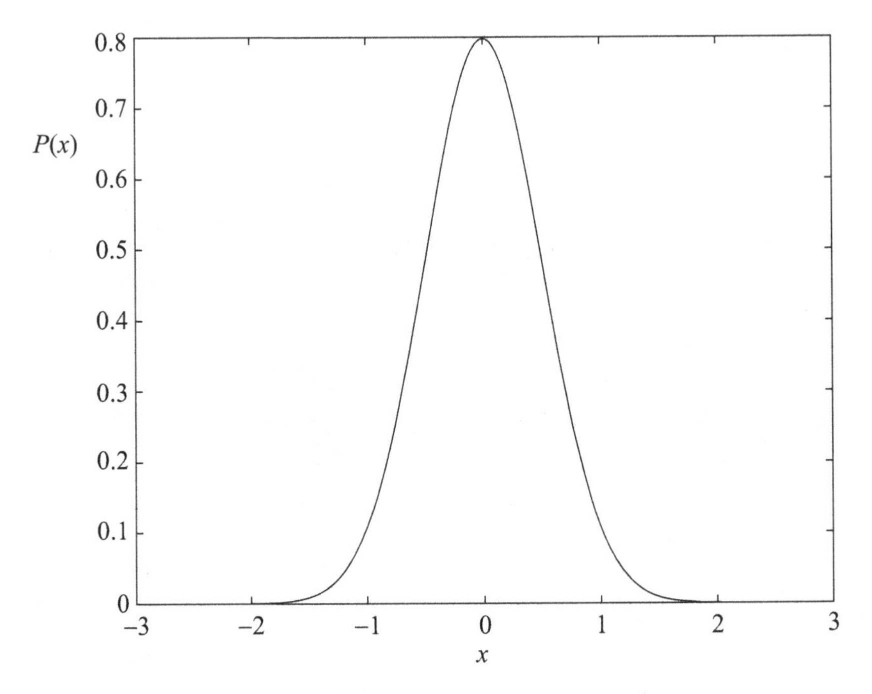

Figure 6.9 The probability density for $\psi(x) = (0.2/\pi)^{1/4}\, e^{-0.1x^2} e^{ikx}$.

$$\mathcal{P}_{\text{all space}} = \left(\frac{0.2}{\pi}\right)^{1/2} \int_{-\infty}^{\infty} e^{-0.2x^2}\, dx = 1.$$

So the probability of finding this particle somewhere in space is indeed 100%.

If you study quantum mechanics, you're likely to encounter other wavefunctions that aren't normalized, and you can generally normalize them by using a procedure much like the one used in this example. If you write the waveform with a multiplicative factor (often called A) out front, you can set the integral over all space of the probability density equal to one and solve for A. The following example shows how to do that for a triangular pulse function.

Example 6.5 *Normalize the triangular-pulse wavefunction in Fig. 6.10.*

The equation for this triangular pulse can be written as

$$\psi(x) = \begin{cases} Ax & 0 \le x \le 0.5, \\ A(1-x) & 0.5 \le x \le 1, \\ 0 & \text{else}, \end{cases}$$

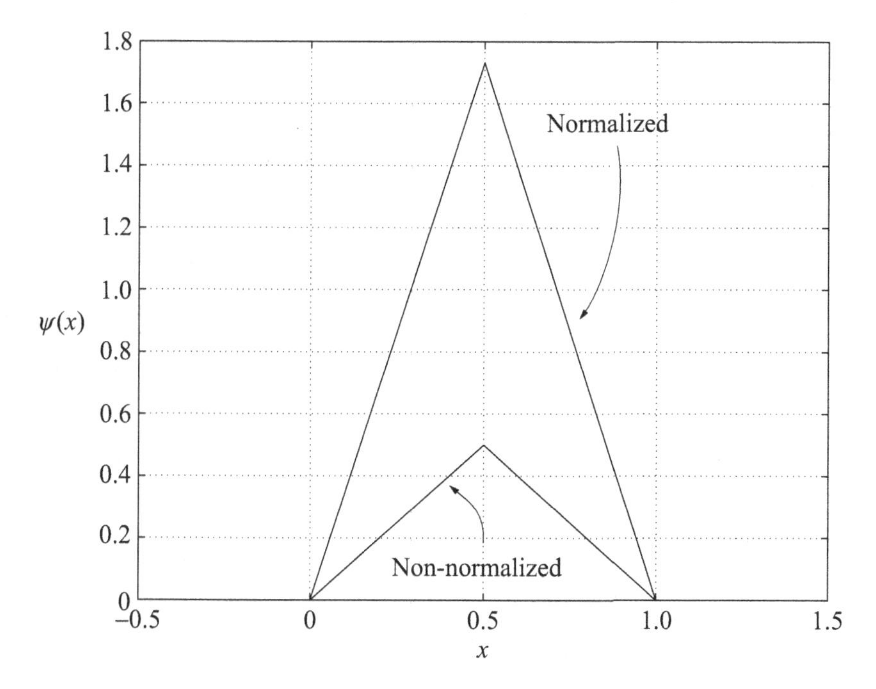

Figure 6.10 Non-normalized and normalized triangular-pulse wavefunction.

which can be plugged into the probability density integral:

$$\mathcal{P}_{\text{all space}} = 1 = \int_{-\infty}^{\infty} \psi^*(x)\psi(x)dx.$$

Thus

$$1 = \int_0^{0.5} (A^*x)(Ax)dx + \int_{0.5}^{1} \left(A^*(1-x)\right)(A(1-x))dx,$$

and pulling out A^*A from each integral leaves

$$1 = A^*A\left(\int_0^{0.5} x^2dx + \int_{0.5}^{1} (1-x)^2dx\right)$$
$$1 = A^*A\left(\frac{1}{24} + \frac{1}{24}\right).$$

In this case, all factors in the equation are real, so $A^*A = A^2$. Solving for the normalization constant A gives

$$A^2 = 12$$

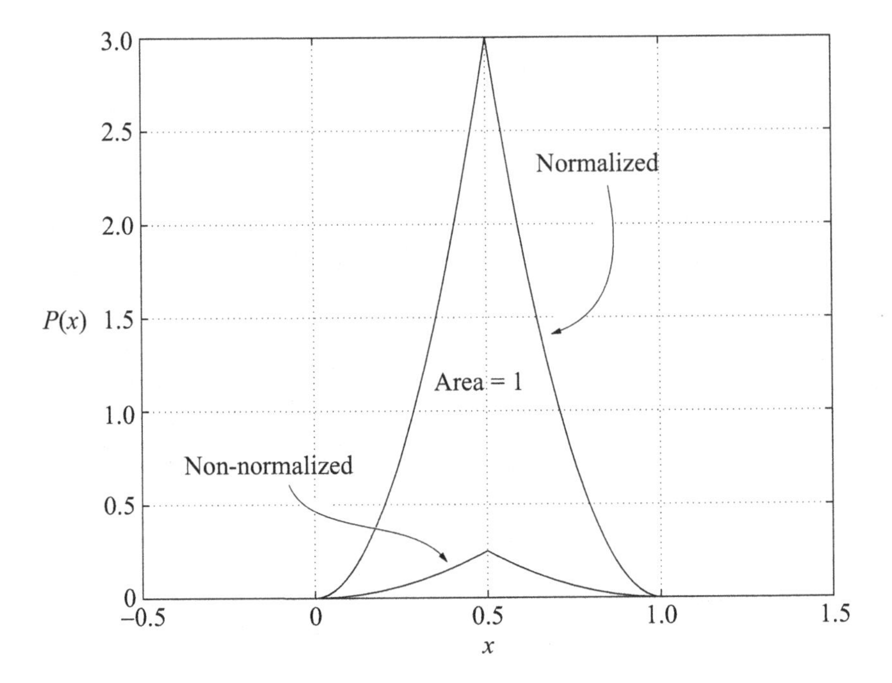

Figure 6.11 Non-normalized and normalized triangular probability density.

and hence

$$A = \sqrt{12}.$$

Figure 6.11 shows the probability density before and after normalization. As desired, the area under the normalized probability density is one, but the shapes of both the normalized wavefunction and the normalized probability density haven't changed from the non-normalized functions; only their scale has changed.

The techniques illustrated in these examples can be used to construct and normalize spatially limited wavefunctions, but it's important that you also understand the wavenumber (and momentum) range of those waveforms. To do that, instead of multiplying an oscillating function by an envelope function, think about the Fourier synthesis approach described in Section 3.3 of Chapter 3. Using this approach, you can construct a spatially limited wavefunction from a single-wavelength function e^{ik_0x} by adding in other single-wavelength functions in just the right proportions to cause the amplitude of the combined function to roll off over distance at the desired rate.

If you attempt to do this using a discrete set of wavefunctions ψ_n and represent the amplitude coefficient of each component waveform as ϕ_n, the combined waveform will be

$$\psi(x) = \sum_n \psi_n = \frac{1}{\sqrt{2\pi}} \sum_n \phi_n e^{ikx}. \tag{6.35}$$

The reason for including the scaling factor of $1/\sqrt{2\pi}$ will be made clear below, but for now you should recall from Chapter 3 that the result of such a discrete sum must be spatially periodic (that is, it must repeat itself over some distance). So, if you hope to construct a wavefunction with a single region of large amplitude, the wavenumber difference between the component wavefunctions must be infinitesimally small, and the discrete sum must become an integral,

$$\psi(x) = \frac{1}{\sqrt{2\pi}} \int_{-\infty}^{\infty} \phi(k) e^{ikx}\, dk, \tag{6.36}$$

in which the discrete coefficients ϕ_n have been replaced by $\phi(k)$, a continuous function that determines the amount of each wavenumber component that gets added to the mix.

If Eq. (6.36) looks familiar, you may be recalling Eq. (3.34) from Chapter 3, which is the equation for the inverse Fourier transform. Thus the spatial wavefunction $\psi(x)$ and the wavenumber function $\phi(k)$ are a Fourier-transform pair. That means you can find the wavenumber function through the forward Fourier transform of $\psi(x)$:[2]

$$\phi(k) = \frac{1}{\sqrt{2\pi}} \int_{-\infty}^{\infty} \psi(x) e^{-ikx}\, dx. \tag{6.37}$$

The Fourier-transform relationship between $\psi(x)$ and $\phi(k)$ has powerful implications. Like all conjugate-variable pairs, these functions obey the uncertainty principle, and that can help you determine the wavenumber content of a given spatial wavefunction.

To see how that works, consider a general Gaussian envelope function of width σ_x. That function can be written as

$$g(x) = e^{-x^2/(2\sigma_x^2)}. \tag{6.38}$$

This is essentially the same envelope function as that of Eq. (6.33), although now the meaning of the constant a is clear: $a = 1/(2\sigma_x^2)$, where σ_x is the standard deviation of the Gaussian wavefunction. Multiplying this envelope function by the interior, single-wavenumber oscillating function $f(x) = e^{ik_0x}$ and normalizing produces the waveform

[2] This is why we included the factor of $1/\sqrt{2\pi}$ in Eq. (6.36).

$$\psi(x) = \left(\frac{1}{\pi\sigma_x^2}\right)^{1/4} e^{-x^2/(2\sigma_x^2)} e^{ik_0x}. \tag{6.39}$$

Now that you know that $\psi(x)$ and $\phi(k)$ are Fourier-transform pairs, you can determine the wavelength content of the time-limited wavefunction by taking the Fourier transform of $\psi(x)$.

That Fourier transform is

$$\begin{aligned}\phi(k) &= \frac{1}{\sqrt{2\pi}}\int_{-\infty}^{\infty} \psi(x)e^{-ikx}\,dx \\ &= \frac{1}{\sqrt{2\pi}}\left(\frac{1}{\pi\sigma_x^2}\right)^{1/4}\int_{-\infty}^{\infty} e^{-x^2/(2\sigma_x^2)} e^{ik_0x} e^{-ikx}\,dx,\end{aligned}$$

so the wavenumber (and momentum) distribution is[3]

$$\phi(k) = \left(\frac{\sigma_x^2}{\pi}\right)^{1/4} e^{(\sigma_x^2/2)(k_0-k)^2}. \tag{6.40}$$

This is a Gaussian distribution around k_0 with width $\sigma_k = 1/\sigma_x$. In other words, it's a distribution with a dominant contribution by k_0 but with other contributing momenta with values near k_0. And, the spread in those values depends on how much the packet spreads in space, as expected.

Specifically, the more quickly the spatial wavefunction rolls off over distance (that is, the smaller you make σ_x), the larger the spread of wavenumbers that must be included in the wavefunction (that is, σ_k must be larger). And, if the range of wavenumbers is larger, the range of momenta must also be larger (since $p = \hbar k$).

Exactly how large is that spread of wavenumber and momentum for a given spread in position? Detailed analysis of the uncertainty in position (which you can find in most comprehensive quantum texts) shows that for a Gaussian wave packet with standard deviation σ_x, the position uncertainty is $\Delta x = \sigma_x/\sqrt{2\pi}$ and the wavenumber uncertainty is $\Delta k = \sigma_k/\sqrt{2\pi}$. And since $\sigma_k = 1/\sigma_x$, the product of the uncertainties in position and wavenumber is

$$\Delta x\,\Delta k = \left(\frac{\sigma_x}{\sqrt{2\pi}}\right)\left(\frac{\sigma_k}{\sqrt{2\pi}}\right) = \left(\frac{\sigma_x}{\sqrt{2\pi}}\right)\left(\frac{1}{\sigma_x\sqrt{2\pi}}\right) = \frac{1}{2}. \tag{6.41}$$

Likewise, the product of the uncertainties in x and p is

$$\begin{aligned}\Delta x\,\Delta p &= \left(\frac{\sigma_x}{\sqrt{2\pi}}\right)\left(\frac{\sigma_p}{\sqrt{2\pi}}\right) = \left(\frac{\sigma_x}{\sqrt{2\pi}}\right)\left(\frac{\hbar\sigma_k}{\sqrt{2\pi}}\right) \\ &= \left(\frac{\sigma_x}{\sqrt{2\pi}}\right)\left(\frac{\hbar}{\sigma_x\sqrt{2\pi}}\right) = \frac{\hbar}{2}.\end{aligned} \tag{6.42}$$

[3] If you have trouble seeing how this integral works, take a look at the chapter-end problems and the solutions on the book's website.

This is known as "Heisenberg's uncertainty principle" and it's a version of the general uncertainty relation between conjugate variables discussed in Chapter 3.

You may hear Heisenberg's uncertainty principle described with words such as "The more precisely you know the position, the less precisely you can know the momentum", and that's true for a certain interpretation of the word "know". If you measure the position of a given particle and then measure its momentum at a later time, you can certainly measure a precise value for each. So there's a better way to think about Heisenberg's uncertainty principle.

Imagine that you have a large number of identical particles, all in the same state (so they all have the same wavefunction). If the spread in positions of this "ensemble" of particles is small, measuring the positions of all of the particles will return very similar values. However, if you measure the momentum of each of these particles, the measured momenta will be very different from one another. That's because particles with small position spread have many contributing momentum states (waves with different wavenumbers), and the measurement process causes the wavefunction to "collapse" randomly to one of those states. Conversely, if the spread in position among the particles is large, then there are few contributing momentum states, and the measured values of momentum will be very similar.

Why do particles in quantum mechanics behave in this strange way? The reason comes back to the wave–particle duality: Because the probability of a particle having a given position or momentum depends on a wavefunction, the relationship between the two will be governed by wave behavior.

The final aspect of quantum waves we'll consider is the time evolution of the wavefunction of a free particle. To do that, we'll have to put the time term back into the spatially limited wavefunction. For reasons discussed above, that wavefunction has a dominant wavenumber k_0 and a range of additional wavenumbers that combine to provide the desired localization. At time $t = 0$, the wavefunction $\Psi(x, 0)$ for a Gaussian wave packet can be written as

$$\Psi(x, 0) = \frac{1}{\pi\sigma_x^2} e^{ik_0x} e^{-x^2/(2\sigma_x^2)}, \tag{6.43}$$

in which σ_x is the standard deviation of the Gaussian position envelope. At time t, this wavefunction is

$$\Psi(x, t) = \frac{1}{\sqrt{2\pi}} \int_{-\infty}^{\infty} \phi(k) e^{i[kx-\omega(k)t]}\, dk, \tag{6.44}$$

in which $\phi(k)$ is the wavenumber function that is the Fourier transform of the position function. In this expression, we've written ω as $\omega(k)$ to remind you that the angular frequency ω depends on the wavenumber (k). So the angular frequency of the dominant wavenumber k_0 is $\omega_0 = \hbar k_0^2/(2m)$, but the angular frequency varies with k as $\omega(k) = \hbar k^2/(2m)$.

Inserting the $\phi(k)$ function for a Gaussian wave packet into the expression for $\Psi(x,t)$ gives

$$\begin{aligned}\Psi(x,t) &= \left(\frac{\sigma_x^2}{4\pi^3}\right)^{1/4} \int_{-\infty}^{\infty} e^{[-(\sigma_x^2/2)(k_0-k)^2]} e^{i[kx-\omega(k)t]}\, dk \\ &= \left(\frac{\sigma_x^2}{4\pi^3}\right)^{1/4} e^{i[k_0x-\omega_0 t]} \left(\frac{\pi}{\sigma_x^2/2 + i\hbar t/(2m)}\right)^{1/2} \\ &\quad \times \exp\left[\frac{-(x-\hbar k_0 t/m)^2}{4\left(\sigma_x^2/2 + i\hbar t/(2m)\right)}\right]. \end{aligned} \tag{6.45}$$

As with the previous integral, if you need help with this one, you can find it in the chapter-end problems and online solutions.

This expression for $\Psi(x,t)$ may not be pleasant to look at, but programs such as Mathematica, MATLAB and Octave can help you explore the behavior of the wavefunction over time. For example, if you choose a particle with the mass of a proton (1.67×10^{-27} kg) moving at a speed of 4 mm/s, the particle's de Broglie wavelength is just under 100 microns. Forming a Gaussian wave packet with a standard deviation of 250 microns results in the wavefunction shown in Fig. 6.12.

In this figure, the wavefunction is shown at times of $t = 0$, 1, and 2 seconds over a spatial interval of about 10 mm. The particle's dominant wavelength is approximately 100 microns, and there are about 2.5 cycles on either side of the central maximum within the standard deviation at time $t = 0$. Each second, the wave packet propagates a distance of 4 mm, so the group velocity of the packet equals the particle's speed, as expected.

The component waves that make up the wave packet all have slightly different phase velocities, but those velocities are about half the group velocity of the packet. To see why that's true, consider that the dispersion relationship for de Broglie waves is

$$\omega(k) = \frac{\hbar}{2m}k^2, \tag{6.46}$$

so the packet's group velocity ($d\omega/dk$) is

$$v_g = \left(\frac{d\omega}{dk}\right) = \frac{\hbar k}{m}, \tag{6.47}$$

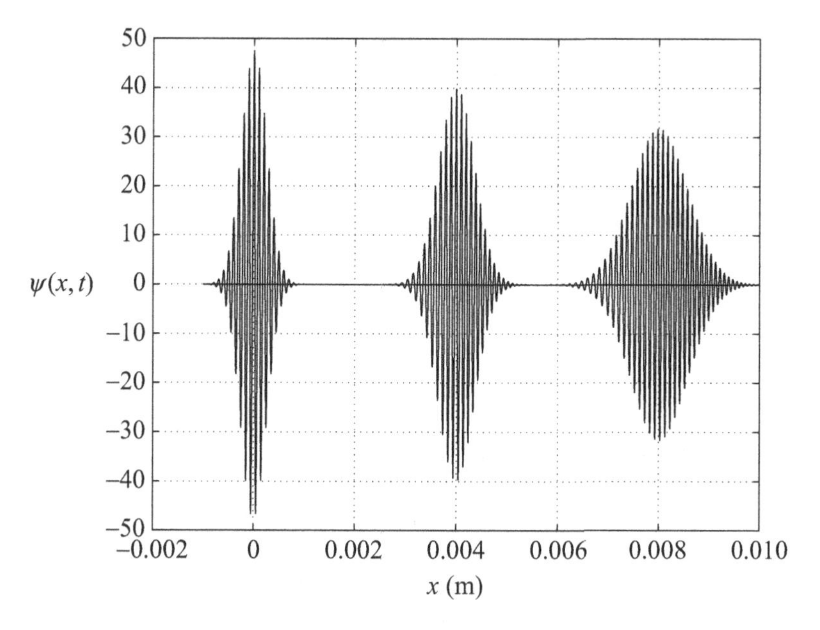

Figure 6.12 The real part of a wave packet at $t = 0$, $t = 1$, and $t = 2$ seconds.

which is the classical particle speed. Recall also that the phase velocity is ω/k, which is

$$v_{\text{p}} = \frac{\omega}{k} = \frac{\hbar k}{2m}. \tag{6.48}$$

This is half the wave packet's group velocity, and half the particle's speed.

Figure 6.13 shows the probability density of the wave packet moving through space as time passes. As you can see in both the plot of $\Psi(x, t)$ and the plot of the probability density, the wave packet isn't just moving, it's also spreading out as time passes. As explained in Section 3.4 of Chapter 3, dispersion occurs whenever the component waves that make up a wave packet travel with different speeds. Given that the quantum dispersion relationship is not linear with respect to k, quantum objects are dispersive – the different speeds of the components of the wave packet cause the packet to spread out in time.

All of the matter waves you've seen in this chapter have been for free particles in regions of constant potential energy (which we set to zero); these all have e^{ikx} as their basis functions. As mentioned in Chapter 4, waves have non-sinusoidal basis functions in an inhomogeneous string (in which the density or tension is not constant through the string). This is also true for matter waves in

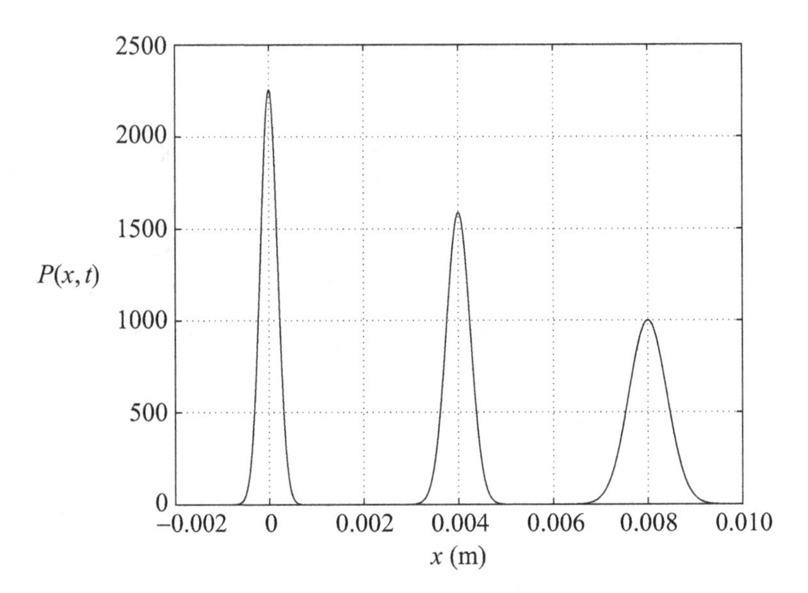

Figure 6.13 The probability density of a wave packet at $t = 0$, $t = 1$, and $t = 2$ seconds.

regions of non-constant potential energy, and you can read about such cases in the "Supplemental material" section on the book's website.

Also on the book's website you'll find the full solution to each of the chapter-end problems, and we strongly recommend that you work through those problems to check your understanding of the concepts and equations in this chapter.

6.6 Problems

6.1. Find the de Broglie wavelength of a water molecule, with mass 2.99×10^{-26} kg, traveling at 640 m/s (a likely speed at room temperature).

6.2. What is the de Broglie wavelength of a proton, with mass 1.67×10^{-27} kg, when it has an energy of 15 MeV?

6.3. The spread in measured positions of an ensemble of electrons is 1 micron. What is the best-case spread in the measured momenta of a similar ensemble?

6.4. Normalize the wavefunction $\psi(x) = xe^{-x^2/2}$ over all space.

6.5. Normalize the wavefunction $\psi(x) = \sin(15x)$ when $0 \leq x \leq \pi/5$, with $\psi(x)$ zero elsewhere.

6.6. Determine the probability of finding the particle of the previous problem between 0.1 and 0.2 meters.

6.7. Show that the wavenumber distribution $\phi(k)$ for the wavefunction in Eq. (6.39) is $\phi(k) = \left(\sigma_x^2/\pi\right)^{1/4} e^{(\sigma_x^2/2)(k_0-k)^2}$.

6.8. (a) Determine the probability of finding a particle with wavenumber between 6.1×10^4 rad/m and 6.3×10^4 rad/m if it has the wavenumber distribution of the previous problem with the values $k_0 = 6.2 \times 10^4$ rad/m and $\sigma_x = 250$ microns.

(b) Compare your answer with the result if $\sigma_x = 400$ microns.

6.9. Show that inserting a Gaussian wave packet (Eq. (6.40)) into Eq. (6.44) leads to the expression for $\Psi(x, t)$ given in Eq. (6.45).

6.10. A very different situation than the free particle is the trapped one: a particle in a potential well. The simplest case is the particle in a box: an infinitely deep, constant potential well as shown below:

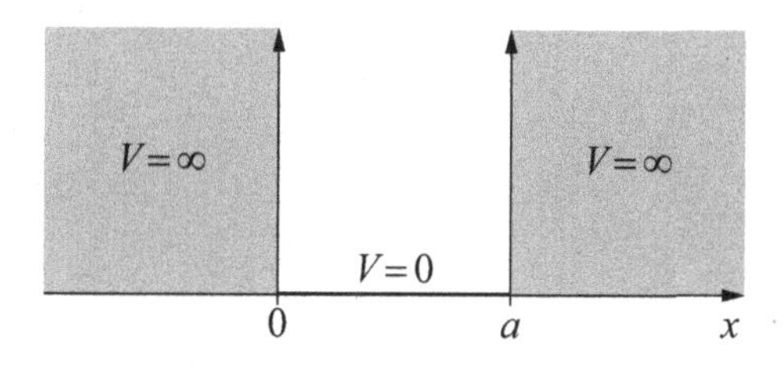

(a) In this case, the wavefunction does not penetrate the side walls and $\psi(0) = \psi(a) = 0$. Show that $\psi(x) = \sin(n\pi x/a)$ satisfies both Eq. (6.21) and the boundary conditions. What values may n have?

(b) Normalize $\psi(x)$.

(c) Plot several wavefunctions with the three smallest n values and compare then with Fig. 3.5 for standing waves on a string.

References

[1] Brigham, E., *The FFT*, Prentice-Hall 1988.
[2] Crawford, F., *Waves*, Berkeley Physics Course, Vol. 3, McGraw-Hill 1968.
[3] Freegarde, T., *Introduction to the Physics of Waves*, Cambridge University Press 2013.
[4] French, A., *Vibrations and Waves*, W. W. Norton 1966.
[5] Griffiths, D., *Introduction to Quantum Mechanics*, Pearson Prentice-Hall 2005.
[6] Hecht, E., *Optics*, Addison-Wesley 2002.
[7] Lorrain, P., Corson, D., and Lorrain, F., *Electromagnetic Fields and Waves*, W. H. Freeman and Company 1988.
[8] Morrison, M., *Understanding Quantum Physics*, Prentice-Hall 1990.
[9] Towne, D., *Wave Phenomena*, Courier Dover Publications 1967.

Index